# Construction 1

# Management
# Finance
# Measurement

Other titles of related interest:

**Building Quantities Explained, fourth edition**
*Ivor H. Seeley*

**Building Technology, fifth edition**
*Ivor H. Seeley*

**Construction 2**
**Environment – Science – Materials – Technology**
*A.V. Hore, J.G. Kehoe, R. McMullan and M.R. Penton*

**Estimating, Tendering and Bidding for Construction**
*Adrian J. Smith*

**Introduction to Building Services, second edition**
*E.F. Curd and C.A. Howard*

**Understanding Hydraulics**
*Les Hamill*

**Understanding Structures**
*Derek Seward*

# Construction 1
# Management
# Finance
# Measurement

A.V. Hore, J.G. Kehoe,
R. McMullan and M.R. Penton

MACMILLAN

First published 1997 by
MACMILLAN PRESS LTD
Houndmills, Basingstoke, Hampshire RG21 6XS
and London
Companies and representatives
throughout the world

ISBN 0–333–64950–8

A catalogue record for this book is available
from the British Library.

Typeset in Great Britain by
Aarontype Limited
Easton, Bristol

Printed in Hong Kong

# Contents

# *Preface*

This book studies the principles and techniques of construction resource management, finance and costing, measurement and pricing. This wide range of topics is of practical use to students and practitioners studying and working in building construction, civil engineering, surveying, planning and development.

The text is intended to help a wide range of students studying construction and built environment topics. The contents will satisfy the principal requirements of courses and self-study for GNVQ/A levels, BTEC awards, HND/HNCs, degrees and professional qualifications.The style of writing is kept simple and supported by clear explanations, a structured layout, practical examples and diagrams. The highlighted definitions, checklists and keyword summaries will also help students preparing for tests, examinations and assignments. The text assumes a minimum of prior knowledge and uses underlying principles to develop an understanding of topics used by professional practitioners in the construction industry.

Some subjects have been distilled into compact treatments which have the advantage of being the first of their type and which will enable everyone to gain a good overview of the construction landscape. However, all the subjects deserve further investigation and we hope that this book will be a starting point for many further studies and careers.

Alan Hore, Joseph Kehoe, Randall McMullan, Michael Penton

# 1  *Resources*

Every construction project is different, but all projects essentially involve the same basic resources. The construction manager is concerned with the management of these resources in order to maximise their use, to minimise their cost and to complete the project within time, within budget, and with safety.

The principal resources involved in a construction project are:

- Materials
- Labour
- Plant
- Finance.

The correct management of these resources ensures that the contractor completes the project according to the contract.

## Materials

The construction manager is responsible for ensuring that a system is in place which manages the procurement of materials. A *materials management system* has been described as having the aim of:

*ensuring that all materials are delivered to site to enable them to be incorporated in the works at the right time, in the correct quantity, at the best cost and of the correct quality.*

In order to meet this aim, good material management should therefore follow the following steps:

- Scheduling
- Requisitioning
- Ordering
- Receiving and handling
- Storage and security
- Issuing
- Incorporating.

**Typical materials**
Cement
Nails
Paint
Concrete
Bricks
Door furniture
Reinforcement
Sanitaryware

## Scheduling

**Procurement officer:**
the person who purchases the
materials on behalf of the
company, usually known
as the buyer.

Using the Bill of Quantities or specifications and drawings the procurement officer will take-off and schedule all the material requirements for the project.

## Requisitioning

When the quantity of materials is known the schedule can be sent out to potential suppliers and a quotation or price requested. The procurement officer must ensure that the potential suppliers are aware of the quantity of the material required and the time at which it is required.

When quotations from different suppliers are received they are compared – not just on cost but also with respect to other factors such as the following:

- Discounts (bulk or otherwise)
- Packaging
- Delivery times
- Delivery charges
- Part load charges
- Guarantees
- Compliance with specification
- Quality.

## Ordering

After consideration of the selection factors, a supplier is chosen and an order is placed; assuming that the contractor has won the contract for the project. The order needs to contain clear details of quantity and the latest delivery dates for the materials. The exact delivery address should be made absolutely clear, as it is not unknown for materials to be delivered to head office, when they are actually required 50 kilometres away at the site!

## Receiving

When delivered to site the material needs to be checked against the order in terms of quantity and quality. To allow this, the site office should receive a copy of the order from head office when the order is placed.

Some time before the order is due, the construction manager should check to make sure that the order is indeed going to arrive at the time expected. Preparation for receiving materials should be made in advance of delivery. For example, labour may have to be available to help off-load the goods and space must be allocated for the materials if they are not to be immediately used.

If the goods are to be directly placed in position then the works must be ready to receive them. For example, cladding panels can be lifted directly from the back of the delivery vehicle by crane and immediately offered up to their final position. Ready-mix concrete suppliers may charge for waiting time if the truck has to wait more than 30 minutes before discharging the concrete.

## Storage and security

The proper storage facilities for materials should be prepared on site. Cement requires a weatherproof store with enough space to allow for the cement to be used in rotation. Valuable items such as electrical goods must be kept in a secure store to avoid damage and theft. To avoid loss of sand or larger aggregates, hard standings may have to be provided. The provision of storage areas is considered when the site layout is prepared and, unless there is good reason to the contrary, the site layout plan should be adhered to.

## Issuing

The issuing of materials to operatives needs to be strictly controlled. A named store person should be in charge of issuing the materials, and their duties will include keeping a record of what has been issued and to whom it has been issued. This aspect of stock control is invaluable in a materials management system, as it prevents excessive wastage of materials as well as theft.

## Incorporating

The incorporation of the correct quantity and quality of material is an important stage. The construction manager must ensure that only properly trained or qualified operatives handle the material and place it in the works. This is particularly important if the materials are of a hazardous nature.

**Some hazardous materials**
Fuel
Paint
Gas cylinders
Concrete

Interwoven with the above steps in material management is the prevention of waste at all times together with the safe handling of the materials. An often-quoted example of the scale of material wastage is that for every 100 houses built there is enough waste of materials to build another 10 houses!

The legal requirements for the Control of Substances Hazardous to Health (COSHH) must be followed, and the provision of safety equipment such as Personal Protective Equipment (PPE) is fundamental.

**Examples of PPE**
Hard hat
Gloves
Goggles
Breathing apparatus
Steel toe-capped boots

# Labour

Labour is the most valuable and sensitive resource that the construction manager deals with. Dealing with the range of labour, from directly-employed staff and operatives to subcontracted staff and operatives, requires the construction manager to have good personnel skills. These skills include the ability to:

**Range of labour**
Bricklayer
Secretary
Planner
Carpenter
Electrician
Subcontract labour
Plumber
Painter
General operative

● Communicate
● Motivate
● Be fair
● Develop team work
● Control
● Organise.

The construction manager must also have up-to-date knowledge of relevant legislation in relation to recruitment, treatment and dismissal of employees.

**Some relevant legislation**
Employment Protection
  (Consolidation) Act 1978
Employment Act 1989
Employment Act 1990
Equal Pay Act 1970
Sex Discrimination Act 1986
Race Relations Act 1976
Health and Safety legislation

A working environment should always have in place *grievance and discipline* procedures in order to deal with any problems, or potential problems, between employer and employee.

The construction manager who works within the framework of the relevant legislation and who implements any agreed procedures goes a long way towards avoiding any disputes which may cause, at worst, disruption to the contract programme and progress and, at best, a disgruntled workforce.

### Labour planning

The labour force required on any project needs to be planned at an early stage. The aim of labour force planning is to establish not only what is to be done but who is to do it. From the overall construction programme, together with the method statement (see Chapter 2), it is possible to forecast the amount, the type and the flow of labour.

The amount of labour on site should be planned to avoid peaks which may require additional supervision and welfare facilities for short periods. This can be achieved, for example, by moving non-critical activities on the programme to attain a more balanced labour pattern.

Even in times of high unemployment certain skilled operatives may not be available and the construction manager must make allowances by either carrying out the particular activity in another way or rescheduling the activity to a time when the labour will be available. Correct forecasting at an early stage avoids problems and prevents delays occurring later on.

In balancing the labour force, care should be taken to build it up in an organised manner, avoiding sudden changes in the numbers of each trade on site. The availability of labour will depend upon the following factors:

- National economy
- Activity in the construction industry
- Training and education
- Technology to be used
- Complexity of the project
- Location of the project.

## Motivation

Motivation of the workforce is an important aspect in achieving an environment of good industrial relations, so the construction manager should strive to ensure that all employees are well motivated. Some of the factors which motivate employees are:

- Good leadership and management
- Pay
- Working conditions
- Team membership
- Feeling that the company cares
- Job security
- Interesting and challenging work
- Good welfare facilities
- Reasonable working hours.

It is in the interests of the project that the workforce is a contented one because, in addition to the individual human benefits, a contented workforce is also a productive workforce.

# Plant

**Typical plant**
Crane
Dumper
Forklift
Scaffolding
Hand tools
Mixer

Like materials management, the management of plant is concerned with ensuring that a system is in place which maximises the use of plant in terms of output and cost. The underlying principle behind *plant management* is to:

*enable construction activities to proceed according to programme by having the appropriate plant available when required.*

Plant management can be broken down into the following stages:

- Selection
- Procurement
- Utilisation
- Control.

## Selection

The type of plant selected for a project depends upon a number of factors including those listed below:

- Site space
- Availability of plant
- Cost
- Method of construction
- Task to be performed
- Ease of operation
- Durability
- Fuel usage
- Effect of weather on usage
- Output
- Safety.

An overriding factor to be considered when making a selection is whether the plant will in fact increase productivity and also whether its use will lead to safer working methods.

## Procurement

Having selected a suitable item of plant for the particular task, the method of obtaining that plant has to be considered. The main procurement options for plant are:

- Hire
- Purchase
- Lease.

When making the decision as to whether to buy, purchase or lease plant, the following factors are relevant:

- Type and quantity of work
- Cost of hiring
- Cost of purchasing
- Cost of leasing
- Tax allowances
- Training of operatives
- Service and maintenance implications
- Storage and security arrangements
- Legal requirement (licence)
- Insurance
- Possibility of generating income (hire out to other contractors).

## Utilisation

How often an item of plant might be used can be worked out from the construction programme. If considering company-wide usage then the programmes for all projects need to be examined. The aim is to obtain maximum usage of plant, together with proper maintenance to avoid breakdowns and to ensure safety.

All plant, whether hired or owned, should be considered in terms of cost per hour. These costs will include the following components:

- Capital cost
- Interest charges
- Insurance
- Licence cost
- Fuel and oil
- Repairs
- Transportation
- Erection and dismantling if applicable
- Storage and workshop costs.

## Control

Control of plant involves the following processes:

- Recording of delivery to site
- Installation of plant on site
- Maintenance and service
- Repairs and overhauls
- Ensuring operator is briefed.

When plant is delivered to site the time and date of delivery should be recorded. The installation period should also be noted, particularly if rates per hour for installation and usage are different.

Registers of any maintenance and service work together with any repairs and overhauls are required for certain items of plant under safety legislation. For example, scaffolding must be inspected by a competent person at least once every seven days, and after bad weather, and a register must be kept of all inspections on a prescribed form.

Only trained operators should be allowed to use items of plant in order to ensure the safety of all operatives on site. Plant utilisation sheets should be used to keep a record of the usage of all plant on site. To minimise cost, hired plant should be put off-hire as soon as it is no longer required, even if the supplier chooses to leave it on site for a day or two after completion.

# Finance

Finance is one of the resources available to the contractor. Control of finance is important as it is affected by everything that the company does. It is also affected by the management of the other resources, such as plant, materials and labour. In order for the company to survive it must at least break even and preferably make a profit.

Later chapters deal with the sources of finance available to the contractor, together with finance forecasting or cash flow forecasting for the company as a whole. At project level there are a number of important steps which can be taken to ensure that the cost of materials, plant and labour is controlled – thereby maximising the return on the finance used.

### Cost control

**Types of budgets**
Company budget
Objectives budget
Capital budget
Operating budget
Flexible budget

Cost control at project level is concerned with making sure that costs do not exceed the budgeted costs.

To help financial control, costs are broken down into the following different cost types:

- Fixed costs
- Variable costs
- Direct costs
- Indirect costs.

---

**Cost control terms**

*Budget*

A predetermined plan expressed in financial terms.

*Cost*

The amount that the contractor has to pay to complete the activity or project.

*Price*

The amount that the contractor has placed in the Bill of Quantities in respect of each item. This amount should be received from the client for carrying out the activity or project.

*Value*

The amount that the contractor will receive from the client. It includes overheads and profit.

*Contract value*

The final amount that the contractor receives from the client, whereas the *contract price* refers to the bid price originally submitted. By its nature the bid price does not include the additional cost of variations and additions to the contract.

*Profit*

The difference between cost and value. Anything which reduces the cost of an activity will add to profit, and vice versa.

---

## Fixed costs

*Fixed costs* can be defined as

*those costs that do not change, irrespective of the level of production.*

In other words, the costs still have to be met even if no output occurs on site. For example, the site manager's salary has to be paid over holiday shut-down periods and therefore it is a fixed cost.

Examples of fixed costs include:

- Head office overheads
- Site management costs
- Site accommodation costs
- Site security costs.

## Variable costs

*Variable costs* can be defined as

*those costs which rise or fall with respect to production.*

Examples of variable costs are:

- Raw materials, bricks, concrete
- Some plant hire cost, excavators, mixers
- Fuel for direct production
- Wages of labour directly involved in production.

## Direct costs

*Direct costs are costs incurred in the actual production of a unit of work or activity.*

For example, a bricklayer's wages plus the cost of bricks and mortar used in the production of a square metre of brickwork is a direct cost to that unit of work.

## Indirect costs

*Indirect costs are costs which cannot be directly contributed in their entirety to a unit of work but are nonetheless incurred.*

For example, the salary of the supervisor or site manager cannot be entirely charged to the cost of the unit of brickwork and is therefore an indirect cost. The classification of costs into different types depends upon how the cost is viewed. The site management cost is a direct cost to the project in its entirety but is an indirect cost to units of work within that project.

# Cost value reconciliation

The control of finance at project level is concerned with collecting cost data and breaking it down so as to compare the data with the budgeted costs.

The budgeted costs are those values included in the Bill of Quantities or other priced document, after deduction of profit.

The actual cost of an activity or unit of work can be determined by using the following type of information source:

- Labour allocation sheets
- Plant hire invoices
- Materials invoices
- Material transfer sheets
- Plant transfer sheets
- Subcontractor invoices
- Daywork sheets.

All the costs incurred in a specific time period are broken down and sorted so as to allocate them to cost centres. Cost centre headings vary from company to company, but the following are common headings:

- Different labour categories
- Plant
- Material
- Site overheads
- Head office overheads.

The headings can conveniently be set out in a table format and allocated beside each activity. The percentage completion of the activity is compared with the same percentage of the value.

For example, if the value less profit in the Bill of Quantities for brickwork to foundations is £9100.00 and the activity is 80 per cent complete then the contractor can expect to be paid £7280.00. If, however, costs are £8000.00 when 80 per cent complete then it alerts management to a potential problem: that the cost of the brickwork to foundations is going to overrun its costs.

By being alerted early to the potential cost overrun, management can take the necessary steps to redeem the situation. However, without an early warning cost control system, such as cost value reconciliation, the relevant cost information may not be available until it is too late.

A thorough understanding of cost types, how to collect cost data and how to compare data is vital to ensure that the contractor manages the financial resource effectively.

A *cashflow forecast* is a prediction of what will happen as a result of a given set of circumstances.

A *budget* is a planned result that a company aims to attain.

# Contracts

The contractor who undertakes a project will enter into a contract with a client. The contract is a legal agreement between the client and contractor and sets out the basis of their relationship throughout the period of the work.

*A **contract** is an agreement between persons giving them rights and duties in law.*

For a contract to be legally valid, the following 'elements' of the contract have to be present:

- Capacity
- Intention to create legal relations
- Offer and acceptance
- Consideration.

## Capacity

Generally, everyone has full legal powers to enter into a contract. There are broad exceptions to this which include infants and minors under the age of eighteen years who cannot contract, except in certain circumstances.

Persons of unsound mind and those unbalanced by intoxication do not have the capacity to enter into contracts. Also, some types of contracts are outside the powers (*ultra vires*) of corporations. Corporations, in other words, cannot enter into contracts unless they have the power to do so.

## Intention to create legal relations

It is required by contract law that the parties actually intend to create legal relations. This is different from social, domestic and family agreements. For example, if you agree to collect your brother or sister at a particular time, this does not constitute a contract.

## Offer and acceptance

The offer to do something must be made and this is different from a mere attempt to negotiate. Tenders for building work constitute an offer. An offer does not remain open indefinitely. There is usually a time period specified as to how long the offer remains open.

An offer to do something must be accepted before a contract can be established. The acceptance must be unconditional and be communicated to the person who made the offer.

## Consideration

This is an essential feature of a contract. The most common forms of consideration are the payment of money and the provision of goods. All parties to the contract must obtain something in return for their promise other than something that they are already entitled to. Consideration must be worth something in the eyes of the law and the courts are not concerned whether the bargain is a good one but simply that there is a bargain.

# Standard forms

Because contracts can be complex, the construction industry has developed standard forms of contract which can be used for building and civil engineering work. A standard form of contract details the contractual terms to be applied between the parties, and it identifies the various documents which represent the legally-binding agreement.

The contract documents may consist of some or all of the following:

- The agreement
- The conditions
- The tender
- The specifications
- The Bill of Quantities
- The drawings.

The form of contract specifies the liabilities, payment methods and risk between the parties in relation to time, cost and quality.

Depending upon how the project procurement is structured, contracts are required between the contractor and client, sub-contractor, and contractor, consultant and contractor, and consultant and client.

Standard forms of contract have the advantage of having been tried and tested over many years and parties to the contract will be familiar with these forms. Standard forms are available

**Procurement systems**
Traditional
Design and build
Construction management
Management contracting

for particular sectors of the construction industry and can be split into the following types:

- Building works
- Civil engineering works
- Government works.

## Forms of contract

The main building works form of contract is published by the *Joint Contracts Tribunal for the standard form of building contract* (JCT); the main civil engineering form is known as the *ICE Contract*.

The full title of the civil engineering contract is 'Conditions of contract and forms of tender, agreement and bond for use in connection with works of civil engineering construction issued by the Institute of Civil Engineers, the Association of Consulting Engineers and the Federation of Civil Engineering Contractors'.

A newer contract titled *New Engineering Contract* (NEC) is also available for use in both the civil engineering and the building sectors.

---

**Standard forms of contract**

Standard form of building contract 1980 edition
  Private with quantities
  Private without quantities
  Private with approximate quantities

  Local authorities with quantities
  Local authorities without quantities
  Local authorities with approximate quantities.

GC Works 1 and 2

**JCT Intermediate form of contract 1984 edition (IFC 84)**

JCT Agreement for Minor Works (MW 80)
JCT Standard Form of Building Contract with Contractor's Design (CD 81)

JCT Management Works Contract (JCT 87)

General Conditions of Government Contracts for Building and Civil Engineering Works (GC/Works/1 and GC/Works/2)
*(continued)*

**Standard forms of contract** (*continued*)

Conditions of Contract and Form of Tender, Agreement and Bond for use in connection with Works of Civil Engineering Construction (ICE)

Conditions of Contract (International) for Works of Civil Engineering Construction (FIDIC)

Association of Consultant Architects Form of Building Agreement (ACA Form)

New Engineering Contract (NEC 1993)

It should be remembered that the standard form of contract can be adapted and altered to suit the particular circumstances.

A construction manager needs a working knowledge and understanding of relevant contract conditions in order to manage the project in terms of time, costs and quality.

**Typical contract clauses JCT 80**

Interpretation, definition
Contractor's obligations
Contract sum
Architect's instructions
Contract documents
Statutory obligations, notices, fees and charges
Levels and setting out the works
Materials, goods and workmanship
Royalties and patent rights
Person-in-charge
Access for architect to works
Clerk of works
Variations and provisional sums
Contract sum
Value added tax
Materials and goods unfixed or off site
Practical completion and defects liability
Partial possession by employer
Assignment and subcontracts
Injury to persons and property and employer's indemnity
Insurance against injury to persons and property
Insurance of the works against clause 22 perils

(*continued*)

**Typical contract clauses JCT 80** (*continued*)

Date of possession, completion and postponement
Damages for non-completion
Extension of time
Loss and expense
Determination by employer
Determination by contractor
Works by employer or persons employed or engaged by employer
Certificates and payments
Finance – tax deduction scheme
Outbreak of hostilities
War damage
Antiquities
Nominated subcontractors
Nominated suppliers
Fluctuations
Contributions, levy and tax fluctuations
Labour and materials cost and tax fluctuations
Use of price adjustment formulas
Settlement of disputes – arbitration

**Keywords for resources**

The following is a list of some keywords used in this chapter. Use the list to test your knowledge and, if necessary, use the text to learn about the terms.

| | | |
|---|---|---|
| Budget | Fixed cost | Requisitioning |
| Capacity | Incorporating | Utilisation |
| Consideration | Motivation | Value |
| Cost | Procurement | Variable cost |

# 2 Organisation and Planning

A construction project must be completed successfully in terms of time, cost, quality and safety. Therefore, the project must be properly organised and planned by the contractor who wishes to undertake the work.

The construction firm or *contractor* first becomes involved in a project when it is invited to tender for a project. The *tender* is an offer from the contractor to carry out the work for a stated sum of money. If the tender offer is accepted by the client, a *contract* is agreed between the client and the contractor.

The planning process is involved from the beginning and can be conveniently separated into the following three phases:

1. Pre-tender planning
2. Pre-contract planning
3. Contract planning.

**Typical projects**
Housing schemes
Shopping centres
Office blocks
Leisure centres
Roads
Bridges

## Planning processes

*Planning concerns the ways in which resources (materials, labour, plant and money) can best be arranged so as to achieve the project objectives.*

The arrangement of the resources will establish the overall plan for the project in constructing terms.

The following factors will affect the project construction plan:

- Time
- Cost
- Quality
- Availability of labour
- Availability of plant
- Type of site
- Construction methods
- Contract.

## Time

The amount of time available for the works to be carried out may be dictated by the client; for example a shopping centre may need to be ready for the Christmas shopping period. Alternatively, the contractor may be asked to indicate how long it anticipates the project will take.

The contractor's estimation of the project duration may be the competitive edge that it has over other contractors tendering for the same project, even though its bid price is higher.

If time is the primary concern of the client, then the contractor who offers the shortest contract period may be awarded the contract.

These effects of time make it important to use a systematic and methodical approach in the planning process. The preliminary assessment of costs is heavily dependent on time and needs to be correctly planned in order to produce a competitive tender.

## Cost

For a contractor to be competitive, the costs of the project need to be minimised without affecting the quality of the work and without affecting the safety of the construction process. Here again planning is of vital importance and involves consideration of the most effective and efficient methods of construction.

## Quality

The quality of the finished project will be determined by the contractor's approach to the control of activities. The *construction plan* will enable the contractor to control activities and avoid unforeseen situations, because the majority of problems will have been anticipated and allowed for.

## Labour

**Trade skills**
Bricklayer
Carpenter
Decorator
Electrician
Plumber

Particular construction activities will require particular skills. Electrical work, for example, will require qualified electricians and bricklaying will require trained bricklayers.

The contractor will need to employ different trades on site at specific times depending upon the nature of the project and will therefore have to know what these times are. This is achieved by converting the project plan into a *programme*.

## Plant

The planning of plant involves similar factors to the planning of labour. Certain items of plant must be available at specific times, otherwise there will be delays and extra costs. But some plant, such as cranes, is expensive and may also be needed on other construction sites. Possible problems are avoided by the use of effective planning.

**Typical plant**
Excavator
Dumper truck
Concrete mixer
Scaffolding
Cranes

## Site

The location of the construction site will determine the construction strategy or plan. If the project is on a very confined site without much room for storage of materials, then the scheduling of material deliveries to the site will be of vital importance and is a key outcome of the planning process.

It may only be possible to deliver materials at a particular time of day or night and this will have to be known well in advance. The price of materials from subcontractors or suppliers may well be increased by special delivery times, such as outside normal working hours.

## Construction methods

A contractor must have a thorough knowledge of construction methods in order to plan and organise properly. Only by knowing how the materials and parts of a building are put together can the contractor effectively sequence the various activities involved in constructing.

## Contract

Restrictions in the contract may affect the planned sequence of activities, particularly if the project is to be completed in phases, or if only certain parts of the site are available at any one time. These contractual restrictions must be allowed for in the planning.

## Personnel

Before looking at the stages and process of planning you need to be aware of the range of expertise needed by the people in the planning team.

The types of personnel involved will obviously depend upon the size, complexity and the anticipated value of the project, together with the nature of the contractor's organisation.

A planning team typically includes the following people:

- Planner
- Contracts manager
- Site manager
- Estimator
- Materials and plant procurer
- Surveyor
- Construction director
- Planning supervisor (safety).

---

**Role of planning supervisor**

- Coordinate the health and safety aspects of project design and planning
- Ensure that the Health and Safety authorities are notified of the project
- Ensure cooperation between designers
- Ensure that a pre-tender stage health and safety plan is prepared
- Advise the client on health and safety matters when requested to do so
- Ensure that a health and safety file is prepared.

---

The grouping of the people in the planning team depends upon the size of the company. For example, in a large construction company there will be a dedicated planning department to carry out the planning function and lead the other members of the team through the planning process.

In a small company the person undertaking the planning role may well be the contracts manager or the site manager. However, in the early planning stages of many larger projects the site manager will not yet have been appointed. The combination of personnel should give the necessary expertise required to allow an effective plan to be produced.

The planner will bring a knowledge of planning and programming techniques (discussed in the next chapter) while the contracts manager and site manager will have a practical and technical input.

The estimator, together with the surveyor, will be able to advise on the cost of different techniques and methods of construction as well as on the labour costs of different plans.

The materials and plant procurer is experienced in obtaining supplies and therefore brings a valuable insight into availability. He can also advise on the output levels of plant and the suitability of different types of plant for a project.

The construction director should ensure that the planning team is working towards its objectives of producing an effective plan, while providing an overall company view of the balance of company resources on different projects.

The responsibility of the team will be to produce the following outcomes:

- An overall construction strategy
- Site layout plan
- A master programme
- Schedule of materials
- Schedule of plant
- Schedule of labour
- Short-term programmes
- Monitoring and control procedures
- Health and safety plan.

These planning outcomes must be produced while also bearing in mind the client's objectives and also ensuring that all activities are carried out in an effective and safe manner.

---

**Health and safety plan**

Provides the health and safety focus for the construction phase of a project.

**Health and safety file**

Records information for the end user and those who might be responsible for the structure in the future. It give details of the risks that need to be managed during maintenance, repair and renovation.

---

# Planning stages

The process of planning can be divided into the following three phases:

- Pre-tender planning
- Pre-contract planning
- Contract planning.

To a certain extent each phase contains similar activities, the difference being in the amount of detail and accuracy required at each stage.

## Pre-tender planning

**Contract documents**
Form of contract
Drawings
Specifications
Bill of Quantities

The time available for pre-tender planning may vary between six and eight weeks, depending upon the type of project. While the contractor knows that he may not win the tender, resources in terms of time and personnel must nonetheless be made available to ensure that a competitive tender is submitted. The contractor depends upon a certain number of tenders being successful in order to stay in business.

During this phase the contractor will decide on the construction strategy and produce a pre-tender report. To achieve this the planning team must study the contract documents and visit the site.

The study of the contract documents should highlight any discrepancies in the document, any onerous or unusual clauses in the contract, any special or unusual construction details, and any unusual item in the specification or Bill of Quantities.

The visit to the site should reveal the following details:

- The exact location of the site
- The access to the site
- Details of services
- Geography of the area
- Local knowledge about the site
- Location of tips in relation to the site
- Local availability of labour
- Local availability of material and plant
- Local weather conditions
- Whether any construction work is commencing or completing in the area
- Condition and closeness of surrounding buildings.

The knowledge acquired from the desk study of the contract documents and from the site locality investigation is combined with the method statement and used to produce a pre-tender report. See later in the chapter for details of method statements. This pre-tender report helps the contractor to establish the risks involved in the project. Adjustments can also be made to the project cost and time estimates and so allow a competitive tender to be submitted.

---

***Pre-tender health and safety plan***

- General description of the work
- Details of timings within the project
- Details of risks to workers
- Information to confirm competence or resource adequacy of the principal contractor
- Information for preparing a health and safety plan for the construction phases
- Information for welfare provision.

---

***Checklist of pre-tender documents***

- Tender summary
- Correspondence file
- Site inspection report
- Method statements
- Outline construction programme
- Outline organisation structure
- Subcontractor lists
- Suppliers quotations
- Cost breakdown
- List of site layout requirements
- Health and safety plan.

## Pre-contract

If the contractor is successful in obtaining the work the process of pre-contract planning can commence.

During this second phase the initial method statements and outline programme are analysed in detail with a view to converting them into a working document which can be used for monitoring and control purposes.

It is at this stage that the timing of activities is set, together with a reappraisal of the sequencing of activities that was put forward at pre-tender stage.

The mechanics for the awarding of contracts to the suppliers and subcontractors is put in place, with the contractor seeking to obtain better terms and conditions now that the works are certain.

A detailed site layout is prepared to show the arrangement of site accommodation, material storage and plant in a manner which will enable the work to be carried out effectively and efficiently.

The site organisation structure should now be formalised, naming the key site personnel and showing the lines of reporting between people and groups (Figure 2.1).

Site services such as water, electricity and telephone can now be confirmed with the relevant bodies and connection dates identified.

Materials which have long delivery periods may need firm ordering at this stage even though the contract commencement date may be some time away.

### Checklist of pre-contract planning documents

- Correspondence file
- Subcontractors' file
- Suppliers' file
- Method statements
- Site layout plan
- Organisation structure
- Master construction programme
- Labour resource schedule
- Material schedule
- Plant schedule
- Health and safety plan.

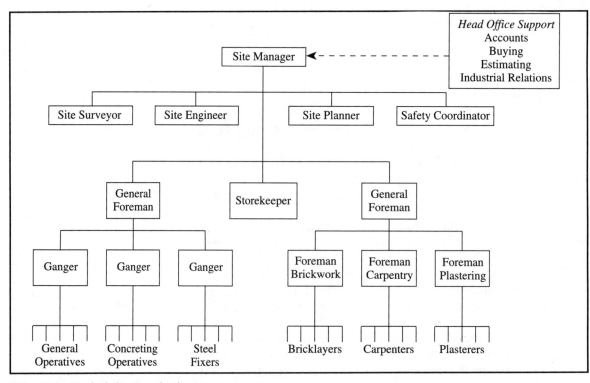

**Figure 2.1**  Typical site organisation structure.

### Contract planning phase

This phase of the planning process takes place during the construction period and involves planning processes which are essentially short term.

The site manager must now break down the master programme into monthly and weekly sub-programmes. Increasing amounts of detail are now required to ensure that activities take place at the correct time and in the correct sequence.

Use will be made of the method statements and programmes to allocate specific tasks to the labour force on a weekly or daily basis. The site manager will issue daily allocation sheets which will list the tasks to be undertaken.

An important aspect of the contract planning phase is the exercise of monitoring and control activities to ensure that the project is running smoothly. These aspects will be discussed in Chapter 3.

---

*Construction health and safety plan*

- Arrangements for ensuring the health and safety of all who may be affected by the construction work
- Arrangements for the management of health and safety of construction work and monitoring of compliance with health and safety
- Information about welfare arrangements.

---

# Site layout

During the pre-contract planning phase there must be considerable focus on the layout of the site. The contractor must ensure that facilities such as accommodation, plant, temporary work areas, site roads and the like are correctly positioned to allow the efficient progress of the works.

Key people involved in the planning of site layout include:

- The site manager
- The planner
- The contracts manager.

The planning team must refer to the tender documents and method statements and take the following factors into account when planning the site layout:

- Construction activities and sequencing
- Efficiency
- Control
- Movement of labour, plant and materials
- Access
- Accommodation
- Health and safety
- Subcontractor requirements
- Public relations
- Statutory conditions.

### Construction activities and sequencing

The order of activities may affect the positioning of accommodation and other facilities in order to avoid conflicts or interference with the progress of the works. During excavation work, for example, space needs to be allowed for the free movement of plant.

It may be possible to arrange early completion of some parts of a project which can be used for storage of materials and equipment. On a housing project, for example, the construction of detached single-storey garages might take place first and then be used to store valuable items of equipment or materials.

### Efficiency

Operations can be streamlined' to give maximum efficiency in use of labour and plant. The avoidance of double-handling of materials can be achieved with a little forethought. An example might be the correct positioning of bricks close to where they can be easily reached by the bricklayer.

Walking distances should be kept to a minimum so that operatives do not have to travel long distances to locate materials, equipment or to use the welfare facilities.

Adequate space for delivery of materials and unloading will minimise disruption and delays and thereby improve efficiency.

### Control

The construction site should be organised so that it is easy to control activities on the site.

The location of the site accommodation will help in the control of personnel entering and leaving the site. For example, raised site offices can give management a clear view of the site at all times.

## Movement

The flow of material, plant and labour around the site can be organised to avoid bottlenecks and waiting times. The movement of plant, such as dumper trucks, can cause hazards and the safety of site operatives must be given high priority when planning the site layout.

## Access

Site access can be considered in two parts:

● Access to the site
● Access within the site.

Access to the site may be limited by road widths, width restrictions, hump-backed bridges and other restrictions which hamper large deliveries. These restrictions will affect the choice of delivery vehicle and may prevent wide, long or heavy loads being delivered.

Access within the site must allow for the safe movement of plant, without endangering the workers on site or damaging parts of the completed structure.

Proper temporary roads with adequate turning circles or through roads may provide a suitable solution for more effective access.

## Accommodation

The contractor must consider the type and amount of accommodation needed, such as site offices, meeting rooms, changing rooms and canteens. The standards for these must meet both legislative and efficiency requirements.

The overall standard of accommodation will depend upon the duration of the contract and the need for human comfort. Facilities such as good daylighting, heating and ventilation should be regarded as an absolute minimum.

**Some legislation**
Health and Safety at
　Work Act
Construction, Design and
　Management Regulations
Control of Substances
　Hazardous to Health

## Temporary services

It may be convenient if temporary services, such as water and electricity, can be brought on to site close to their final position.

Placing toilet facilities close to existing sewers may save on temporary pipe runs, but this needs to be balanced against the need to have toilets close to the works to avoid long travel distances for the operatives.

Avoidance of any temporary overhead cables such as electrical or telephone is a basic rule.

## Security

The security of the site must be seen as important for two reasons: to prevent the possibility of theft, and to prevent unauthorised people from being on the site. Trespassers, particularly children, may injure themselves on site and the contractor may be blamed.

The choice of security arrangements will generally depend upon the location of the site and the threat of vandalism. Hoarding around the site and the use of security personnel may be a minimum requirement in some areas, while fencing may be adequate in others.

## Health and safety

The contractor will have a duty to ensure that the site has been laid out with respect to health and safety matters such as those outlined here.

The provision of temporary fire escapes during the construction phase may need to be considered. Dangerous substances, such as flammable liquids, need to be properly identified and given safe storage.

The erection and positioning of scaffolding must be planned to avoid accidents or damage to the scaffolding from moving plant.

Any dangerous existing services, such as a gas pipe, need to be identified and areas which pose a threat to health and safety may need to be sectioned off.

## Subcontractors' requirements

Due attention to the needs of the subcontractors will prevent conflict or misunderstandings during the construction phase. Accommodation, positions for specialist equipment and adequate working space may all be required.

## Public relations

Considerate treatment of the general public and of the neighbours adjoining the site will go a long way towards achieving a harmonious project.

Careful positioning of noisy plant away from neighbours will help prevent complaints and possible stoppages of work. Noise pollution is controlled by legislation and the fitting of screens or noise dampers to machinery may also be required. Dust and debris should not be allowed to escape from the site and the boundaries should be kept tidy.

# Method statements

During the pre-tender and pre-contract planning phases the contractor will have produced a method statement for the works. Because the method statement is central to the planning process it is described here as a special item:

A **method statement** *is a comprehensive description of the contractor's approach to carrying out the construction work.*

A method statement usually includes the following items:

- The location of the site
- The nature of the site
- The contractor's expertise for the type of work
- The contractor's intended timescales for the works
- The intended order of the works
- The amount and type of labour required
- The amount and type of plant required
- An assessment of the output for the various activities.

The layout and format of method statements can vary from company to company. Some companies prefer to list all the elements in a report format, while others like to fill in a ready-made form or proforma. See Figure 2.2.

Whatever layout is used, the contractor must ensure that all the activities or operations associated with the project are included in the method statement. The planner must use a systematic analysis of all activities in order to produce the 'programme' of works.

The early method statement, at the pre-tender phase, will include alternative methods for carrying out the activities so as to allow choice. The later method statement, during the pre-contract stage, will have a more detailed analysis.

The headings included in a typical method statement comprise the following:

**Method statement headings**
Operation number
Operation
Quantity
Method
Plant
Labour
Safety features
Output
Comments

| METHOD STATEMENT | | | | | | | | |
|---|---|---|---|---|---|---|---|---|

Project:_____    Date:_____ Sheet: _____

Client:_____    Prepared by:_____

| Operation Number | Operation | Method | Materials Quantity | Plant | Labour | Safety | Output | Comments |
|---|---|---|---|---|---|---|---|---|
| | | | | | | | | |

**Figure 2.2**   Method statement.

## Operation number

This is simply a number given to a particular operation in order to identify it. The operation number can subsequently be used in the contract programme.

# Operation

This represents a particular operation such as excavation works, external walls or roof works. The operation or activity usually consists of a single trade or parcel of work.

# Method

This entails a detailed written description of the contractor's approach to the operation, how the stages of the operation will be carried out and in what order.

# Quantity

The amount of materials to be used in connection with the operation will be included: for example, the quantity of bricks required for the external walls or the amount of material excavated from trenches during excavation work.

# Plant

The type of plant and equipment to be used for the operation will be listed here. Careful consideration will be given to the size, capacity and usefulness of the plant for the operation.

# Labour

The amount of labour required will be listed and the particular trades identified: for example, banksman, carpenter, electrician and general labourer. This will help the production of the labour schedules and histograms used later on in the planning process.

**Site hazards**
Objects left on ground
Overhead cables
Falling tools
Moving vehicles
Unprotected holes

# Safety features

If specialist equipment or temporary works are needed they can be noted here. The inclusion of this heading in the method statement will focus the contractor's mind on the safety aspect of each operation. Even small items of safety equipment should be included, such as personal protective equipment, as well as more specialist items like breathing apparatus or temporary works which give protection to the operatives.

## Output

An estimate of the rate of output, that is, the time taken to carry out the activity, should be included here. This estimate will help the estimator to build up the tender figure as well as help the planner to carry out detailed programming later on.

## Comments

Here the contractor will include any comments which are pertinent to the operation. Comments may identify which operations are subcontracted or if certain materials or plant need to be ordered early owing to their specialist nature. These comments will be of particular use to the planner and site manager later on in the contract.

---

### Keywords for organisation and planning

The following is a list of some keywords used in this chapter. Use the list to test your knowledge and, if necessary, consult the text to learn about the terms.

| | | |
|---|---|---|
| Access | Layout | Pre-tender |
| Construction plan | Master programme | Programme |
| Contractor | Method statement | Sequencing |
| Efficiency | Personnel | |
| Health and safety plan | Planning supervisor | |

---

# 3   *Programming*

All construction work must be programmed to ensure that activities are sequenced in the correct order.

*A **programme** is the conversion of the construction plan to a time-related chart.*

The construction programme sets out the progression and sequence of the works for key activities in sufficient detail to show the contractor's plan for carrying out the works with respect to time.

A programme is important because it supports the following actions:

- Inform the client of the contractor's intentions
- Provide a timetable of activities
- Determine plant requirement time
- Determine materials requirement time
- Determine labour requirement time
- Provide a basis for control and monitoring of activities
- Aid the production of a cashflow chart
- Provide time evidence in relation to claims
- Enable all parties to assess the consequences of delays and interruptions
- Allow an assessment of changes or alterations to the works during the construction period with respect to time and therefore cost.

## Classification

Programmes can be separated into three distinct types. The following classification depends upon the amount of detail included and the timescale of the programme:

- Master programme
- Section programme
- Operation programme.

### Master programme

**Major activities**
Site set-up
Substructure
Superstructure
Fitting out
External works

The master programme includes all the major activities for the entire project. This programme gives a broad picture of the scope of the works and the intended time period. The time period is usually shown in months and weeks and the major activities are plotted against these periods.

### Section programme

The section programme deals with a particular phase or section of the works over a particular time period of the contract. For example, the substructure works can sometimes usefully be considered as a section programme, particularly if the works consist of complicated activities, such as piling or basement construction.

   The time period is again in months and weeks, but a more detailed breakdown of the activities allows the site manager to plan activities from week to week. A section programme can also be used to depict subcontractor's work on site. The section programme is a sub-programme of the master programme and amplifies the sub-programme.

### Operations programme

An operations programme provides a very detailed sequencing of activities for a particular activity or operation. Finishes such as painting can be shown on an operations programme. Each area of the building and the sequencing of these areas can be shown. The operations programme has time periods of weeks and days which allow, for example, the delay period between undercoating paintwork and finishing coats to be clearly shown.

## Techniques of programming

Programmes can be worked out and presented in a number of forms, each with its own advantages and disadvantages. The method of producing the programme is known as the **programming technique** and there are essentially three techniques with which you should be familiar:

- Bar or Gantt charts
- Networks
- Line of balance.

Before studying how to produce any of the three techniques, it is important to know the advantages and disadvantages of each technique and in what circumstances each might be used.

## Bar or Gantt charts

The bar or Gantt chart was developed by Henry L. Gantt, a management scientist who lived during the early part of this century. He developed a pictorial method of showing planned progress against actual progress.

The bar chart shows the time, usually in months and weeks, horizontally along the top, with the activities listed vertically down the left-hand side. See Figure 3.1. The bar chart is the most commonly used programming technique and it has a number of distinct advantages.

### *Advantages of the bar chart*

- Simple to produce
- Easily understood by everyone on site
- Gives a good pictorial representation of the construction sequence
- Can be updated easily
- Different levels of programme can be easily related to each other, for example master, section and operations programme
- Key dates can be easily shown on the programme
- Can be easily used to indicate progress and thereby aid monitoring and control of the project
- Can be drafted on a preprinted proforma.

### *Disadvantages of the bar chart*

- Difficult to show the interrelationships between activities
- The representation of the construction sequence may give too simple a picture.

For example, if an activity is delayed it may be difficult to assess what effect this delay will have on future activities.

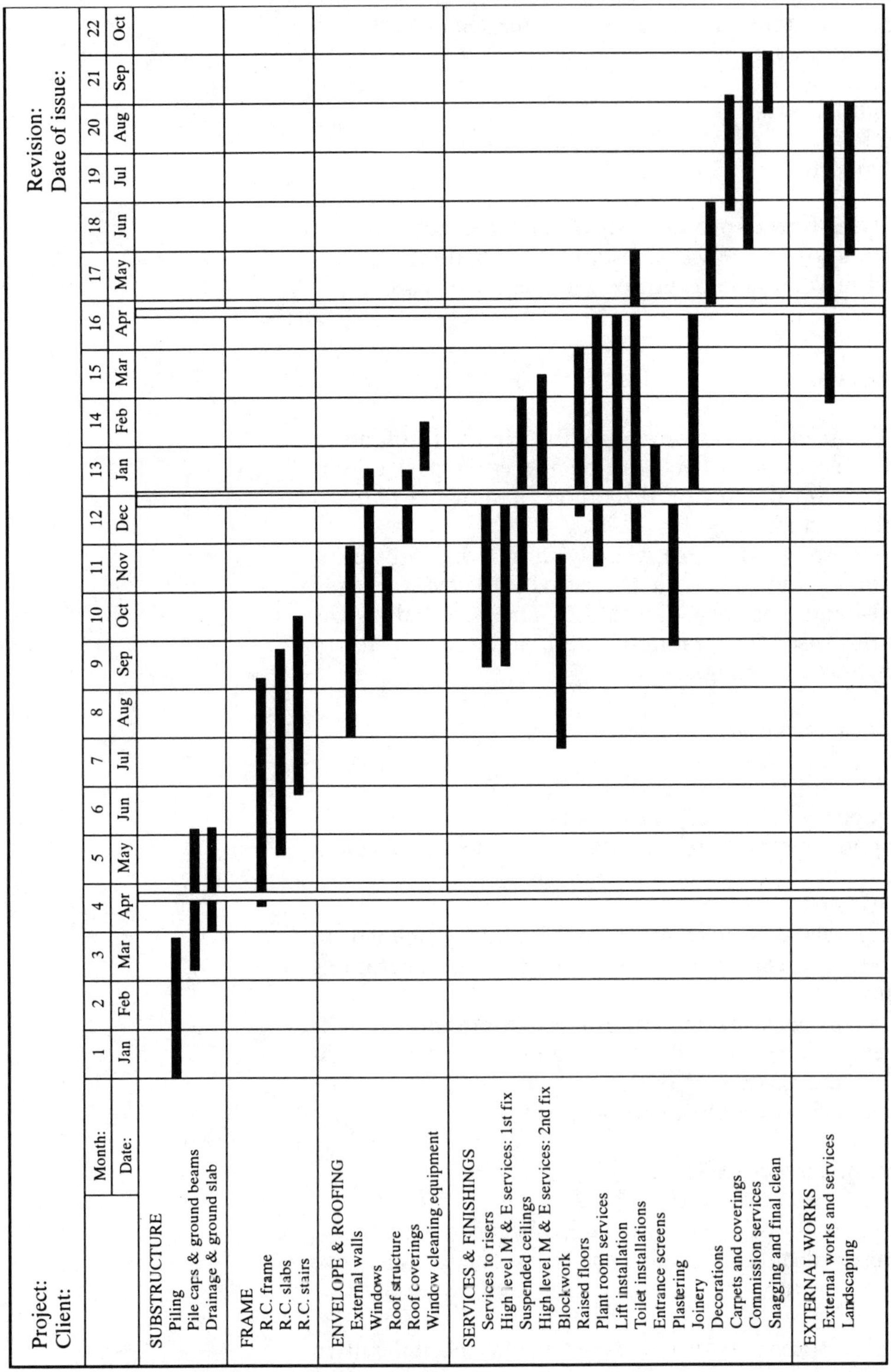


**Figure 3.1** Typical master programme.

Important information which should always be shown on a programme includes the following:

- Name of project
- Name of client
- Contractor's name
- Date of production of programme
- Date of any amendments
- Name of person who produced the programme.

## Networks

This programme can be shown as a network of interrelated activities, highlighting the start and finish times of each activity. There are numerous network programming techniques, but the following two types are traditionally used in the construction industry:

- Activity-on-the-arrow
- Activity-on-the-node (also known as Precedence networks).

The objective of network programming is to identify the *critical* activities and show their relationships. This enables the planner to arrive at a minimum time in which the project can be undertaken.

Network programming may also be known by the following terms:

- Network analysis
- Critical path analysis (CPA)
- Critical path method (CPM).

Network programming has a number of distinct advantages over bar charts and these are listed below.

### *Advantages of network programming*

- Shows the interrelationships of activities
- Requires the planner to think logically
- Critical activities are identified
- Non-critical activities are identified and this allows the planner to balance resource requirements
- Resources, such as materials and plant, which may act as a restraint on the programme and cause delay, are highlighted
- The effect on future activities of a delay on a previous activity can be easily analysed

- Effects of delays can be identified and claims by the contractor for extensions of time to the project can be substantiated more easily
- The effect on the programme of speeding up activities can be assessed
- Networks are suitable for complex and difficult projects.

### Disadvantages of network programming

- Not easily understood by all participants to the contract
- Updating and redrafting can be time consuming and difficult
- Requires the planner and users to have an expert knowledge of the construction processes and methods used in the project.

The use of computers can now reduce the amount of time required to produce and redraft a network. After the network is analysed it can be converted into a bar chart which is more easily understood. The resultant bar chart has the advantages of both techniques. The interrelationships between activities can be shown by linking activities on the bar chart.

## Line of balance

**Examples of repetitive work**
Housing projects
Floors in multi-storey buildings
Tunnelling work
Roadworks
Piling operations
Pipelines

**Gang:** *A group of people involved in a particular trade or operation*

**Trades**
Carpentry
Plumbing
Bricklaying
Electrical
Painting and decorating

The *line of balance* (LOB), which is also known as the *elemental trend analysis* (ETA), is used to programme repetitive work.

The use of bar charts or network analysis for repetitive activities becomes cumbersome and does not balance the rate of work against similar sequential activities.

The line of balance is similar in concept to the bar chart in that bars represent the activities (see Figure 3.1). But in this case the times are plotted on the horizontal axis with *the number of units requiring similar activities* plotted on the left-hand vertical axis.

The concept is of gangs for particular trades moving from one unit to the next unit in the sequence.

Typical applications of this technique are for multiple housing projects or the fitting-out of a number of floors in a multi-storey building. Indeed, any project which consists of a number of similar units or sections with similar activities can be programmed using the line of balance technique.

### Advantages of line of balance

- Rate of working of one trade against another is clearly shown
- Information is easily understood

- Provides good pictorial representation of construction sequences
- Effects of delays are clearly shown
- Resources can be levelled against one another
- Progress can be recorded easily as an aid to monitoring and control
- Requirement dates for materials and plant can be easily identified
- Different rates of output are evident
- Effects of accelerating the activities are clearly shown.

### Disadvantages of line of balance

- Suitable only for repetitive work
- May give too simple a picture of the construction process as a lot of information is summarised on the chart.

# Production of programmes

All the programme types identified in the previous sections involve the same early steps in production:

- List all activities
- Sequence all activities
- Time all activities
- Produce the programme.

## Activities

All the activities in a project can be identified with the aid of the *method statement* which has been prepared earlier in the planning stage. The activities should correlate with particular trades or sections of the work. For example, excavation work requires labourers, and brickwork to damp-proof-course level requires bricklayers. Roof work can be split into two: roof timbers require carpenters and roof coverings require roofers.

## Sequence

When the activities have all been listed, the planner then sequences them into a logical and practical order. This involves the production of a logic diagram, as shown in the 'activity-on-the-arrow' illustration (Figure 3.2).

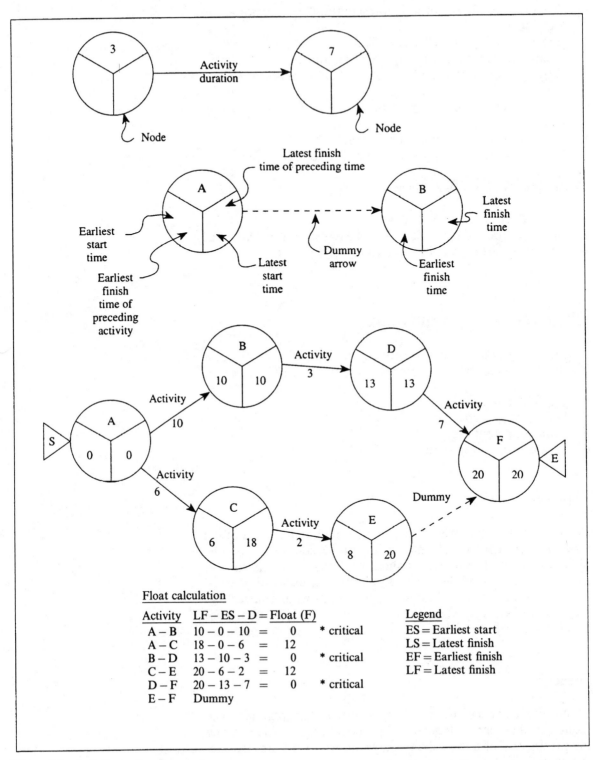

**Figure 3.2**    Activity-on-the-arrow codes.

A broad sequence of activities will include the following:

- Site set-up
- Site clearance
- Substructure
- Superstructure
- Fit-out
- External works.

However, each of these broad areas has to be broken down into activities or operations as explained previously.

There are three fundamental questions that the planner must ask when deciding how and where to place an activity in the sequence. They are:

1. What activities must be completed before this activity can be carried out?
2. What activities can only be carried out after completing this activity?
3. What activities can take place at the same time as this activity?

## Timing

The duration of an activity can be calculated in a number of ways. Reference may be made to published standard times for activities, such as the rate of output for a particular size of concrete mixer, or the number of bricks that a bricklayer can lay in an average eight-hour day.

The planner, together with the estimator, uses the method statement to calculate outputs, while making any allowances for variations in the activity or difficulties that may be encountered with respect to the particular project.

The use of work study principles (as discussed in Chapter 4) may also be appropriate in calculating activity durations. Past experience of particular activities and methods of construction also plays a part in predicting activity durations.

## Programme

If the programme is of the bar chart type, the activities can now be plotted. The resultant draft programme is discussed with the contracts manager and site manager, and their input is used to make amendments.

The site manager who implements the programme on site must believe in, and be committed to, the timescale and sequence construction. The programme is of little use if the site manager believes that he has been set an impossible timescale and feels no ownership of the programme.

### Network production

As previously stated, two types of networks are typically used in the construction industry:

- Activity-on-the-arrow (AOA)
- Activity-on-the-node (AON) or Precedence diagram.

### Activity-on-the-arrow production

Before looking at the mechanics of producing an AOA network, the particular 'code' of networks must be known. Refer to Figure 3.2.

The circle represents an *event* or *node*. It can be divided into three parts, two of which contain time and one which contains part of the activity reference. See Figure 3.2.

- The *arrow* represents an activity. The length of the arrow is **not** related to time.

- The activity description is always written on top of the arrow.

- A broken or dotted arrow is a *dummy*.

A dummy activity is used purely to maintain logic in the network and because it is **not** an activity it does not have a duration. The use of a dummy also avoids two activities having the same reference.

- The duration of the activity is always written below the arrow.

In activity on-the-arrow, the activity *always* has two reference points. For example activity 3–7, or activity A–B.

All projects must begin and end with an activity. All activities between the beginning and the end must be *tied-in* to other activities. This avoids leaving activities 'dangling' or hanging unresolved.

> ### *Explanation of activity-on-the-arrow diagram (refer to Figure 3.2)*
>
> - The figure in the left-hand side of the circle (event) is the earliest start time of the activity on the arrow following the circle. It is *also* the earliest finish time of the activity on the arrow before the circle.
> - The figure in the right-hand side of the circle is the latest start time of the activity on the arrow following the circle. It is *also* the latest finish time of the activity on the arrow before the circle.
> - The earliest start time represents the earliest possible time that an activity can begin and the latest start time is the latest time that an activity can begin.
> - The earliest finish time and latest finish time are the earliest and latest possible times when an activity can end.
> - If the earliest and latest start times are the same for a particular activity, and the earliest and latest finish times are the same for that activity, then the activity is a *critical activity*.

- A *critical activity* is an activity which has least or no 'float time'.
- *Float time* is the amount of spare time available.
- The *critical path* is the pathway or sequence of activities through the network which has the least float time, or no float time. The duration of this pathway is the shortest possible time in which the project can be completed.

A non-critical activity has float, and therefore there is some discretion as to when it is carried out, provided that it is carried out within the timeframe allowed. That frame is the time between the earliest possible start time and the latest possible finish time.

For example, an activity such as drainage works may have an earliest possible start time of day ten. The activity requires twelve days to complete and the latest possible finish is day fifty-five. Therefore, the timeframe available for the drainage works to be carried out is forty-five days. That is, fifty-five minus ten. As drainage takes twelve days, the amount of spare time or float is equal to thirty-three days. The formula for calculating float for a particular activity is as follows:

Latest finish time – Earliest start time – Duration = Float

**Float calculation formula**

$LF - ES - D = F$

Knowing the timeframe and float of an activity allows the contractor to carry out that activity when resources are available. This helps the contractor to balance the use of resources such as labour and plant.

### Activity-on-the-node production

The activity-on-the-node, or precedence, diagram uses similar logic to the activity-on-the-arrow diagram, but it is presented differently. With this technique, the activity is represented by a box or node, with the arrows showing logic between the boxes. See Figure 3.3.

---

**Explanation of activity-on-the-node (precedence) diagram**

- The earliest start time is placed in the top left corner of the node
- The latest start time is placed in the bottom left corner of the node
- The earliest finish time is placed in the top right corner of the node
- The latest finish time is placed in the bottom right corner of the node
- The activity reference is written in the centre of the node, with the activity duration written below the reference.

---

If there is a delay or lag between one activity and another, then this delay figure is written under the arrow. A lag between one activity and another could be, for example, the delay in striking formwork from an *in situ* concrete column, thereby allowing the concrete time to set.

The relationship between nodes can therefore be clearly shown using the following relationships:

- Start to finish
- Finish to start
- Lag start to finish
- Lag finish to start
- Lag start to start
- Lag finish to finish.

The above features of activity-on-the-node (precedence) diagrams allow activities to overlap and this is a distinct advantage over the activity-on-the-arrow network, where it is difficult to show activity overlaps.

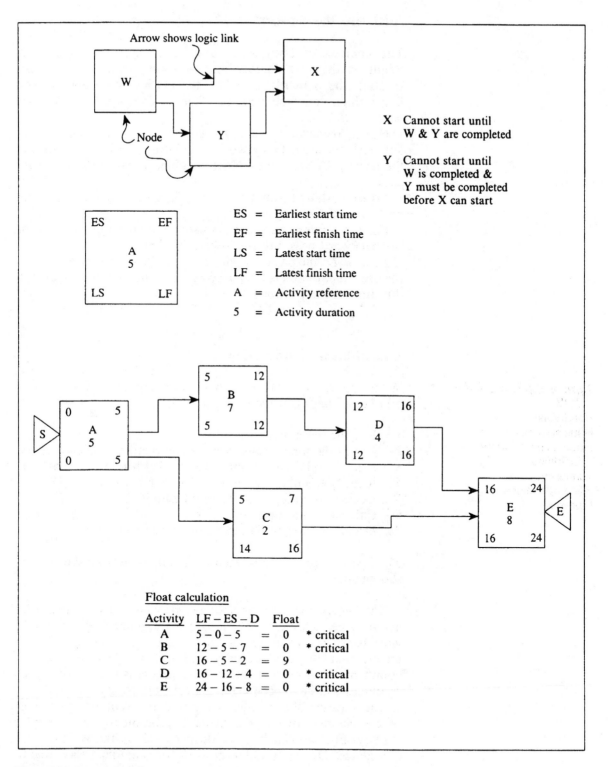

**Figure 3.3** Precedence diagram.

## Analysing the network

The analysis of both AOA and AON networks is similar. Begin at the first activity and, working from left to right through the network, calculate the earliest start and earliest finish times for each activity. This is known as the *forward pass*.

Having reached the end of the last activity, work from right to left back through the network calculating the latest finish and the latest start time for each activity. This is called the *backward pass*.

When working through the network, take care to calculate all paths into the event or node.

For the forward pass, always use the highest number and for the backward pass, use the lowest number.

*Remember*: the pathway with the least float time or no float is the critical path and is the pathway which determines the overall duration of the project.

## Line of balance production

*Typical applications for LOB*

Roadworks
Housing projects
Floors in multi-storey buildings
Tunnelling
Piling operations
Pipelines

When preparing a line of balance programme there are a number of steps to follow (Figure 3.4):

1.  List the main activities.
2.  Assess the time needed for the completion of each activity.
3.  Assess the labour or gang size needed for each activity.
4.  Identify the plant to be used for each activity.
5.  Decide on buffer times between activities.
6.  Calculate total time for one unit.
7.  Decide on rate of handover of subsequent units.

The first four items can be identified and taken from the method statement.

The *buffer* time between activities provides spare time in case activities get delayed or take longer than anticipated. The buffer will absorb delays in completion of a previous activity, which means that the next activity will not be delayed and can commence on time as planned. This feature is important, as the resources will have been arranged for the activity.

The duration of the buffer is normally based on the experience of the planners or site managers in relation to the particular activity. Factors which cause delays, such as the weather, are considered. During foundation work for example, delays due to the weather are quite common.

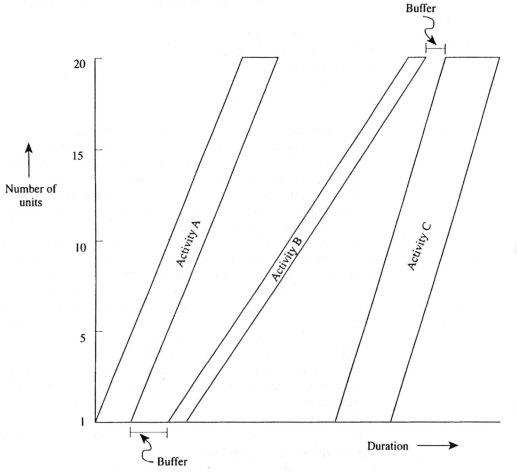

Activity A is proceeding at a faster rate than activity B. Therefore the buffer is at the bottom.
Activity B is proceeding at a slower rate than activity C. Therefore the buffer is at the top.
If the buffer between B and C was placed at the bottom, activity C would crash into activity B.
Output depends upon:   · Type of plant used
                       · Size of gang employed
                       · Nature of activity
                       · Handover rate.

**Figure 3.4** Line of balance.

The slope of the bar or line representing the activity depends upon the rate of working or output. The steeper the line, the faster the activity takes place over the units. Ideally, all the lines should be parallel; however, in practice this does not occur, as different activities proceed at different rates. Excavation for foundations takes place at a faster rate than, say, the brickwork to damp-proof-course level.

The rate of output will depend upon the following factors:

- Type of plant used
- Size of gang employed
- Nature of the activity
- Weather
- Handover rate.

The handover rate may be dictated by the client; for example, the client may want a certain number of houses ready for sale every week. Once the first house is completed, the rest may have to be handed over at a rate of say four per week.

When calculating the rate of output, the planner computes the number of gangs required for a particular activity, bearing in mind the handover rate.

In order to meet the handover rate the number of gangs may be computed to be three and a half. This is known as the *theoretical gang size*. The planner then has to adjust the number of gangs to either three or four, as half a gang is not practical. If three gangs are chosen, the activity line will proceed at a slower rate than the ideal to meet the handover rate and if four gangs are chosen it will proceed at a faster rate.

When the rate of output for all the activities has been calculated with respect to gang sizes and the buffer times have been decided upon, they can be plotted to produce the programme.

---

**Keywords for programming**

The following is a list of some keywords used in this chapter. Use the list to test your knowledge and, if necessary, use the text to learn about the terms.

| | | |
|---|---|---|
| Activity-on-the-arrow | Float | Network |
| Buffer | Gantt chart | Operation |
| Critical activity | Line of balance | Precedence |
| Critical path | Logic diagram | Programme |

---

# 4  *Work Study*

Construction managers are continually trying to improve existing methods and develop new methods for carrying out construction work. The science of *work study* can help them, by providing a systematic approach to the analysis and evaluation of working methods.

Work study techniques have been developing since the early 1900s and they were initially used in the study of factory processes. The aim was to improve productivity, and increase the output from each process. The techniques of work study can equally be applied to the various activities and working methods found on construction sites.

Work study is concerned with two key factors:

- **Methods:** how the activity is carried out
- **Time:** how long it takes to carry out an activity.

The study of these two factors can help bring about better methods and shorter times, with improvements in the following areas:

- Productivity
- Safety
- Cost
- Efficiency
- Human fatigue
- Morale
- Motivation
- Wastage.

An improvement in any of the above areas means that the contractor is more competitive. The more competitive the contractor, the greater the likelihood of winning tenders and obtaining jobs, and therefore remaining in business.

For the worker or *operative*, the benefits can include a safer working method and a less tiring method. If the operatives are aware that the improvement is for their own benefit, as well as for the company, then they may be better motivated.

The critical factors of method and time are also used to divide the topic of work study into two parts:

- Method study
- Work measurement.

This chapter examines each of these parts, along with their associated techniques.

# Method study

Method study is concerned with the way in which work is carried out and in trying to find the most efficient method. When building a brick wall, for example, it is more efficient to have the stack of bricks near where the wall is being built rather than bring the bricks from a distant part of the site whenever they are required.

However, method study looks at more than just the position and location of the bricks. It can also look at how the brick is held, how it is handled, and how it is laid by the bricklayer.

In order to have a systematic approach to studying activities or operations the following steps should be followed:

**Process**
Selecting
Recording
Examining
Developing
Installing
Maintaining

- Select the operation
- Record the operation
- Examine the operation
- Develop the operation
- Install the new operation
- Maintain the new operation.

## Selecting

The first step is to select the operation which is to be studied. This selection may be based on a number of features of a project, such as the following:

- Expensive
- Time consuming
- Dangerous
- Wasteful of resources
- Awkward
- Complicated.

If the operation is a one-off and the contractor is certain that it will not be repeated on that particular project or other projects, then it may not be cost-effective to carry out a full study.

However, the majority of construction activities, when broken down into smaller parts, are used over and over again. Because of this repetition, any improvements will pay dividends in the long term.

## Recording

When the operation to be studied is selected the method or process by which it is carried out must be recorded. To help in recording the activity a number of techniques, such as the following, may be used:

- Outline process charts
- Flow process charts
- Flow diagrams
- Multiple activity charts
- String diagrams
- Templates or models.

The symbols used in the techniques of method study are those recommended by the American Society of Mechanical Engineers and are known as the ASME symbols. Use of the symbols simplifies the recording process and each symbol represents an activity. The symbols are internationally recognised and are shown in Table 4.1.

**Table 4.1**  ASME symbols for work study

| Symbol | Activity | Meaning |
|--------|----------|---------|
| ◯ | Operation | Represents the main stages in a process |
| ▢ | Inspection | Represents an inspection for quality or quantity |
| ▽ | Permanent storage | Represents deliberate storage |
| ◗ | Delay | Represents a delay in the sequence of activities |
| ⇨ | Transport | Represents movement |

### Outline process charts

*An **outline process chart** is a chart showing an overall view of the main sequences involved in a process.*

This is a useful chart which records the order in which the work is carried out. A simple outline process chart is shown in Figure 4.1.

The outline process chart, as its name suggests, only records the outline of the process and uses only two of the ASME symbols: operation and inspection. If it appears that the process is inefficient then further, more detailed studies may need to be undertaken using a flow process chart.

### Flow process chart

*A **flow process chart** is a chart showing the sequence of all activities involved in the process.*

A flow process chart shows greater detail than the outline process chart and is usually prepared using all of the standard ASME symbols. The flow process chart can be one of the following three types:

- A labour flow process chart
- A material flow process chart
- A plant flow process chart.

It may be seen from the example flow process chart (Figure 4.2) that travel, distance and time are included where appropriate. The totals for each symbol are shown at the bottom of the chart. If the chart shows many symbols for either delay or transport, then it highlights possible problems with the process.

The flow process chart can be further enhanced in value by hatching or colouring similar symbols. This allows the more common symbols to be seen at a glance.

The flow of work can be recorded in terms of what happens to the material, labour or plant and equipment. For clarity, however, each resource should be recorded separately.

### Flow diagram

A diagram may be used to give a pictorial representation of an activity. This flow diagram shows the flow of materials, labour or plant in relation to the work environment (see Figure 4.3). A flow diagram is usually used in combination with the flow process chart for the same operations.

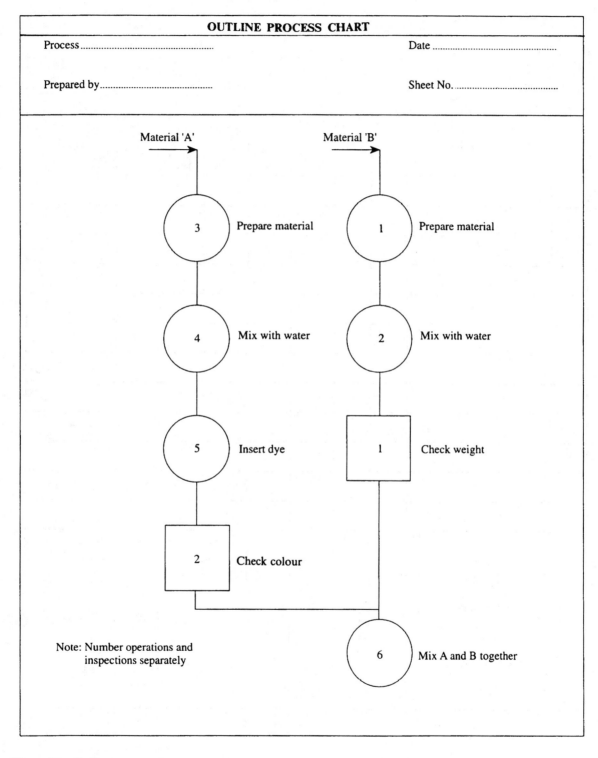

**Figure 4.1** Outline process chart.

**FLOW PROCESS CHART**

Process.............................................    Date.................................................

Prepared by.....................................    Sheet No.......................................

| No. | Operation description | Dist. | Time | Symbol | | | | | Remarks |
|---|---|---|---|---|---|---|---|---|---|
| | | | | ◯ | ⇨ | D | □ | ▽ | |
| | | | | | | | | | |
| | | | | | | | | | |
| | | | | | | | | | |
| | | | | | | | | | |
| | | | | | | | | | |
| | | | | | | | | | |
| | | | | | | | | | |
| | | | | | | | | | |
| | | | | | | | | | |
| | | | | | | | | | |
| | | | | | | | | | |
| | | | | | | | | | |
| | | | | | | | | | |
| ◯ = Operation | | | | | | | | | |
| ⇨ = Transport | | | | | | | | | |
| D = Delay | | | | | | | | | |
| □ = Inspection | | | | | | | | | |
| ▽ = Storage | | | | | | | | | |
| Totals | | | | | | | | | |

**Figure 4.2** Flow process chart.

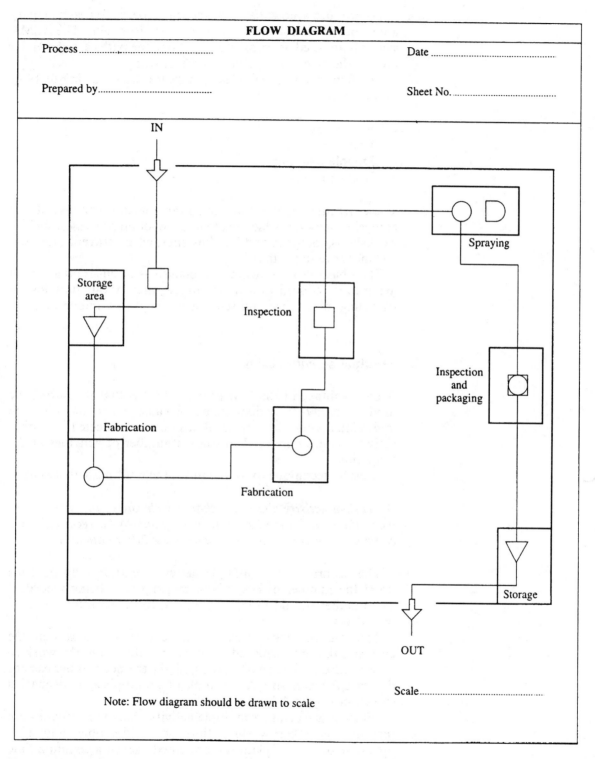

**Figure 4.3** Flow diagram.

To produce a flow diagram, the travel paths for each activity are recorded on a scale drawing of the environment. The ASME symbols are used as before, but this time the paths are shown by linking the symbols together on the drawing.

The flow diagram will clearly indicate the following features of a project:

- Bottlenecks
- Long pathways
- Double-handling
- Back-tracking.

These features make flow diagrams particularly useful for analysing activities which relate to production processes, such as workshops, factories and building sites, where travel routes and distances are important.

The objective of a flow chart analysis is to ensure that travel distance is reduced to a minimum, so that labour time lost in travelling, plant time and fuel use are kept to a minimum.

### Multiple activity charts

A disadvantage of the flow process chart is that it can only be used to study individual items. Because most activities are interrelated (one activity is almost always affected by other activities), you may need to study a number of activities at the same time.

A useful technique for doing this is the multiple activity chart:

*A **multiple activity chart** is a chart which allows the activities of more than one operative or item of plant to be recorded on a common timescale in order to show their interrelationship.*

The accuracy of a multiple activity chart usually depends upon the number of operatives, or gang size, being recorded. The number should usually be limited to between four and six operatives.

The multiple activity chart can be used to check that the correct balance is achieved between operatives and the work to be carried out. The chart may highlight the need to increase or decrease the gang size. An example of a multiple activity chart is shown in Figure 4.4.

When constructing a multiple activity chart the activities are recorded as either working time or as idle time, and each operative or item of plant is represented against a common time scale.

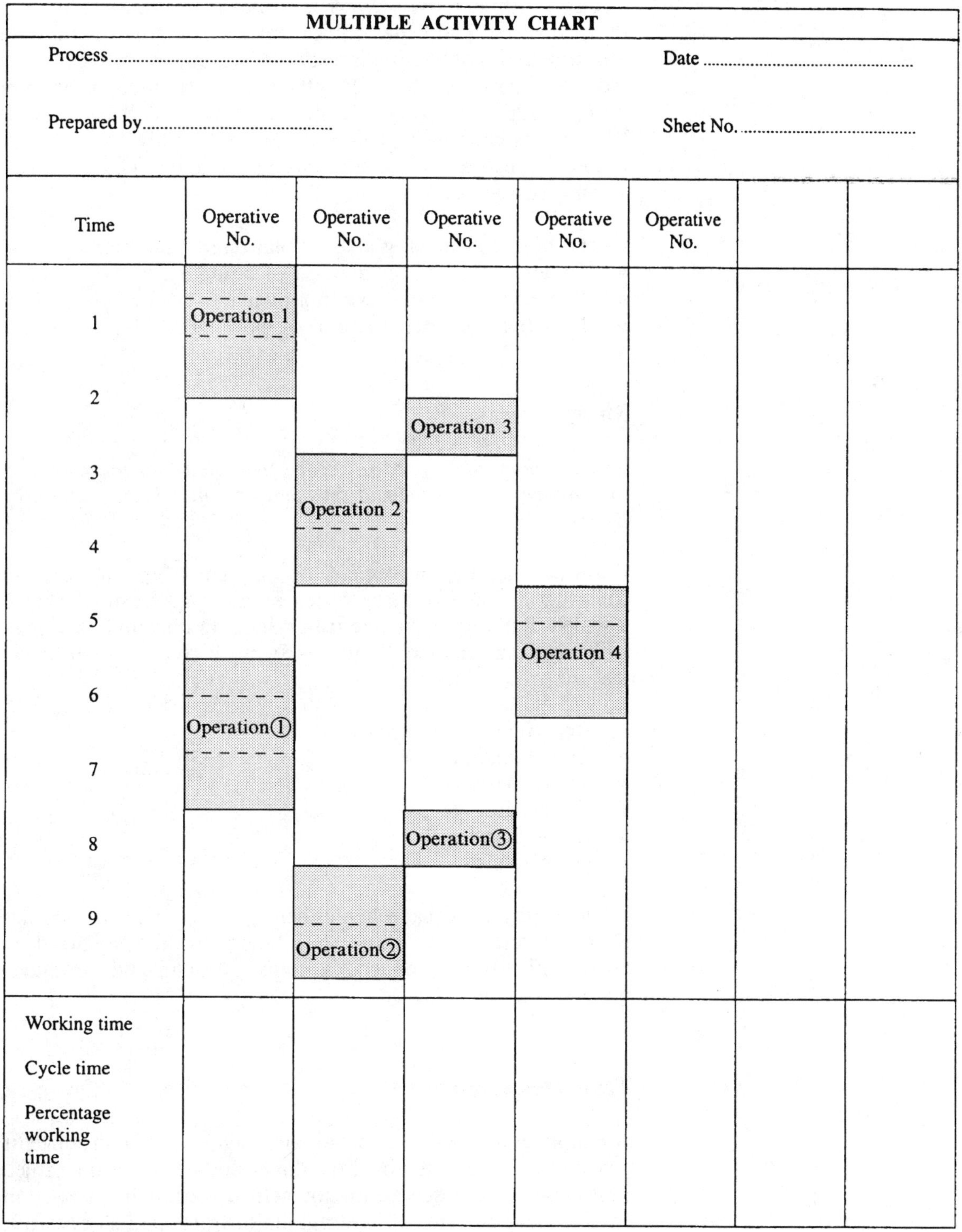

| | | | | | | | |
|---|---|---|---|---|---|---|---|
| **MULTIPLE ACTIVITY CHART** | | | | | | | |

Process.................................

Date.................................

Prepared by.................................

Sheet No.................................

| Time | Operative No. | Operative No. | Operative No. | Operative No. | Operative No. | | |
|---|---|---|---|---|---|---|---|
| 1 | Operation 1 | | | | | | |
| 2 | | | Operation 3 | | | | |
| 3 | | Operation 2 | | | | | |
| 4 | | | | | | | |
| 5 | | | | Operation 4 | | | |
| 6 | Operation① | | | | | | |
| 7 | | | | | | | |
| 8 | | | Operation③ | | | | |
| 9 | | Operation② | | | | | |
| Working time | | | | | | | |
| Cycle time | | | | | | | |
| Percentage working time | | | | | | | |

**Figure 4.4** Multiple activity chart.

The multiple activity chart is useful for organising work so that expensive plant and equipment are idle for the minimum duration and that waiting time and delays for operatives are also reduced. The times for each activity or operation can be found by carrying out a work measurement which will be discussed later in this chapter in the section on time study.

The objectives of the multiple activity chart can be summarised as follows:

- To balance the gang against plant and equipment
- To achieve a quicker method of doing the work
- To decrease the idle, waiting or inactive time
- To achieve a better method of working.

### String diagram

*A **string diagram** is used to record the amount of movement and the pattern of movement of an operative, materials or plant and equipment.*

For example, the movement of an item of plant around a construction site can be recorded by plotting the movement on a scaled drawing of the site using drawing pins and string.

String diagrams (see Figure 4.5) can highlight the following features of a project:

- Repetitive movements
- Travel distances
- Travel paths
- Congestion
- Inefficiency
- Obstructions.

The principle is that when either labour or plant is moving it is not necessarily working and being productive. So if the movement can be reduced a more efficient and productive method may be achieved.

### Templates and models

A template or model of a production layout can be produced to give a two-dimensional or three-dimensional view of the project. This type of representation can help to rationalise a working environment in terms of lighting, major travel paths or relationships between different levels of a process.

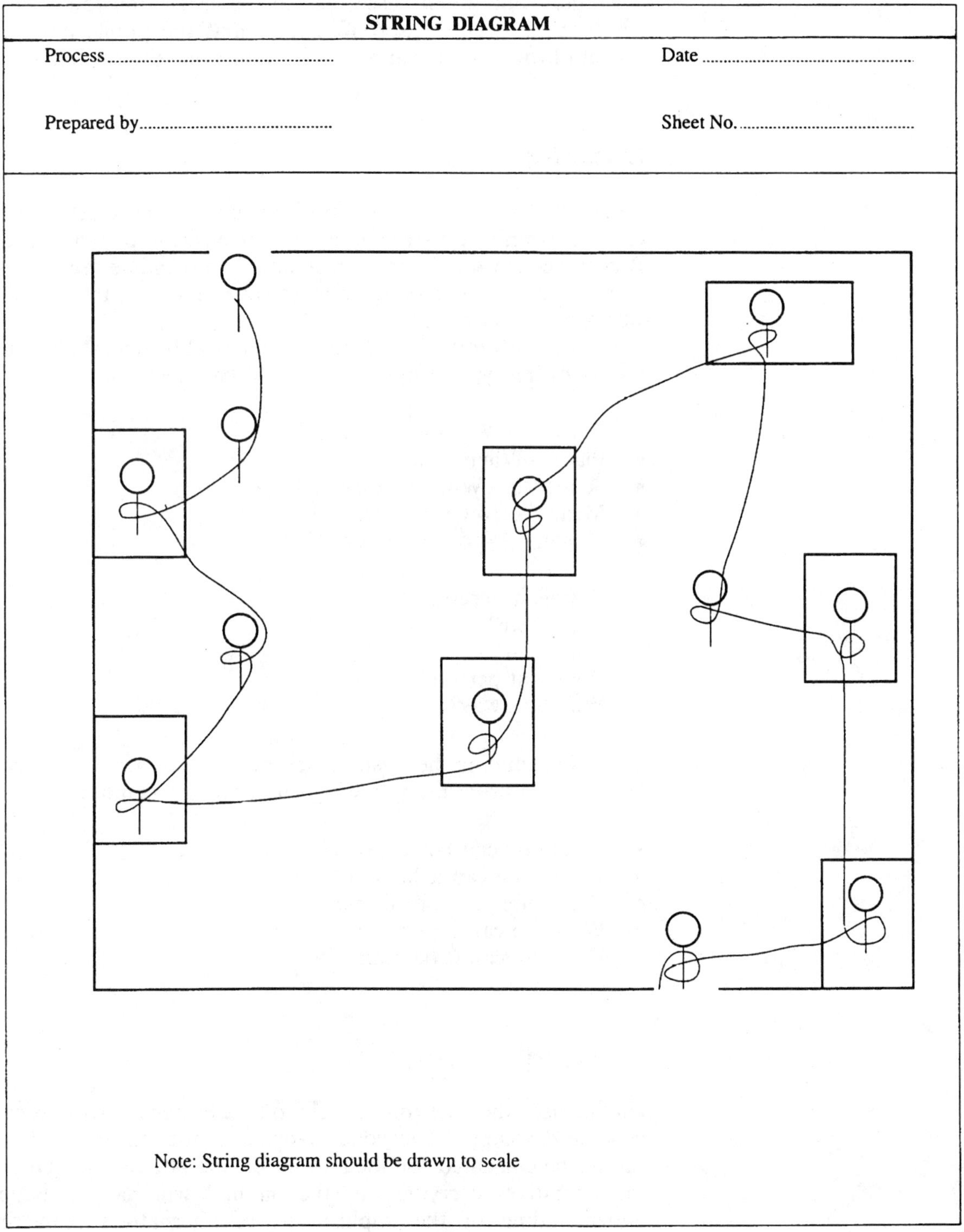

**STRING DIAGRAM**

Process.................................................... Date ...................................................

Prepared by........................................... Sheet No.....................................

Note: String diagram should be drawn to scale

**Figure 4.5** String diagram.

Models are often used in the construction and built environment sectors to show clients and the public how the layout of town centres and large complex buildings will appear.

## Examining

After you have recorded the process that you have selected, it can be examined to see if any improvement can be made. Remember that the improvement should streamline the process so as to make it less dangerous, to save time, and to minimise wastage of resources.

A critical examination of the process must be undertaken and the following type of questions should be asked:

- Purpose – What is achieved?
- Place – Where is it carried out?
- Sequence – When is it carried out?
- Means – How is it carried out?
- Person – Who carries it out?

- Why is it necessary?
- Why there?
- Why then?
- Why that person?
- Why that way?

After examining the existing method, alternative methods can be proposed using the following questioning technique:

- What else can be achieved?
- Where else can it be done?
- When else can it be done?
- Who else can do it?
- How else can it be done?

## Developing

Having used the questioning technique, alternative methods can now be developed. Everyone involved in the existing method should be consulted for their views on the alternatives. Often, the operatives carrying out the method will have a better understanding of the implications of alternatives than the person carrying out the study; involvement of the operatives is crucial to the success of the new and improved method.

## Installing

Once a more efficient and safer method has been developed it needs to be implemented or installed. This can only be achieved by ensuring that the operatives are aware of the improved method and the reasons for changing the existing method. Many new methods fail because of management's inability to explain the new method clearly.

The installation of the new method should be supervised so that any questions arising from its installation can be answered immediately. The installation should be properly planned and arranged. It may be desirable to have a rehearsal until everyone is familiar with the new method.

## Maintaining

Once the new method or process is implemented, it then needs to be maintained. This maintenance may mean that some adjustments and amendments will need to be made to the process, particularly if there are changes to any circumstances related to the new method. The new method should therefore be reviewed relatively frequently, so that it can be fine tuned. Over a longer period of time, the results of the new method should also be reviewed to check that the new method does indeed improve productivity and safety.

# Work measurement

Work measurement is the second part of work study and it is concerned with the time that it takes to carry out an activity.

*Work measurement is the use of special techniques to establish a standard time for a qualified worker to carry out a specified job at a defined level of performance.*

The purpose of work measurement is to find out how long it takes to carry out an activity so that the output from an operative, or a machine, can be measured. In doing so, ineffective time can be highlighted and possibly eliminated.

As with method study there is a disciplined approach to work measurement. A standard approach is to use the following steps:

**Process**
Selecting
Recording
Examining
Measuring
Compiling
Defining

- Select the activity
- Record the activity

- Examine the activity
- Measure the time
- Compile the standard time
- Define the activity.

The first three stages in the process of work measurement are similar to those of method study. The last three stages are concerned with measuring the existing time for the activity and defining a new, more efficient time.

Work measurement can be broken down into the following two aspects:

- Direct work measurement
- Indirect work measurement.

Direct work measurement involves measuring the activity in the field, while indirect work measurement can be carried out at the desk using data which already exists.

There are a number of techniques which can be used to help measure the time of an activity. The principal techniques are:

- Time study
- Activity sampling
- Synthesis
- Analytical estimating.

## Time study

*Time study is a technique for recording the times and rates of working for the elements of a specified job carried out under specified conditions, and for analysing the data so as to determine the time necessary for carrying out the job at a defined level of performance.*

This technique involves observing a competent operative working at a specified activity. Each stage or element of the activity needs to be recorded, timed and rated. The aim is to establish the time for doing the activity at a defined rate under specified conditions.

### Basic process of time study

1. Select work to be measured.
2. Analyse the work cycle and break it down into elements which will be easy to measure.
3. Time and rate each element. This should be done several times if necessary.

4. Extend observed time by calculating to basic time.
5. Average the basic times and add allowances to build up standard times.

Standard time study sheets are available to enable the person carrying out the measurement to record and analyse accurately.

---

*Some important terms used in time studies*

*Element* – a particular part of an activity which is easily defined and measured.
*Rating* – adjustments made to take into account variation in the operative's pace.
*Cumulative timing* – time recorded on time study sheet by stop watch.
*Observed time* – time taken to perform each element of work.
*Standard rating* – the average rate at which a qualified operative will work, given sufficient motivation and instruction.
*Basic time* – time required for carrying out an element of work at standard rating.
*Relaxation allowances* – time allowed for operative to attend to personal needs and to recover from fatigue or tiredness.
*Contingency allowance* – time allowed for certain activities during work; for example, preparation or tidying up such as a painter cleaning brushes.

---

## Time study worked example

An extract from a time study which was carried out on a bricklaying operation for a single-storey domestic dwelling is given in Table 4.2.

**Table 4.2**  Time study table

| | Element | Rating | Cumulative time (standard minutes) |
|---|---|---|---|
| A | Discharge materials | 85 | 0.27 |
| B | Mixing mortar | 75 | 2.30 |
| C | Laying brickwork | 100 | 2.85 |
| D | Pointing brickwork | 90 | 5.13 |

If the total allowances are 28%, 34%, 30% and 33% for elements A, B, C and D respectively, calculate the standard time for each element and for the whole operation (in standard minutes).

### Solution to worked example

| Element | (a) Rating | Cumulative time | (b) Ordinary time | (a × b) Basic time | Allowances (%) | Standard time |
|---------|--------|-----------------|-----------|------------|------------|----------|
| A | 85  | 0.27 | 0.27 | 0.23 | 28 | 0.30 |
| B | 75  | 2.30 | 2.03 | 1.53 | 34 | 2.05 |
| C | 100 | 2.85 | 0.55 | 0.55 | 30 | 0.72 |
| D | 90  | 5.12 | 2.27 | 2.05 | 33 | 2.73 |

Standard time in minutes    5.80

## Activity sampling

*Activity sampling is a technique in which a large number of observations of either labour or plant are made over a period of time.*

On many sites it may be difficult to establish which activities are causing problems in relation to delays or inefficiency. Because the majority of activities are interrelated, it is not an easy task to pinpoint the problems.

One quick and relatively easy technique that can be used to highlight inefficiency, is to carry out an activity sampling exercise. This basically involves spot-checking activities over a period of time to check what is occurring.

Each observation records what is happening at that particular instant and the percentage of productive time is related to the percentage of idle time.

A typical example on a construction site is to check the efficient use of plant. On a prepared sheet (see Figure 4.6), each item of plant is listed against a timescale and it is recorded whether or not the item of plant is being used at particular times. If some items of plant have a high percentage of utilisation, or a high percentage of under-utilisation, then further investigation can be carried out using some of the other techniques discussed earlier in this chapter.

| ACTIVITY SAMPLING SHEET | | | | | | | |
|---|---|---|---|---|---|---|---|
| Site.......................................... | | | | Date.................................. | | | |
| Observer................................ | | | | Weather.............................. | | | |
| Time | Crane | Dumper | Mixer | Hoist | | | |
| 8.00 | ✓ | X | X | X | | | |
| 8.30 | ✓ | ✓ | X | ✓ | | | |
| 9.00 | ✓ | ✓ | X | ✓ | | | |
| 9.30 | X | ✓ | ✓ | ✓ | | | |
| 10.00 | X | ✓ | ✓ | X | | | |
| etc. | | | | | | | |
| Total observations | | | | | | | |
| Percentage working | | | | | | | |

**Figure 4.6** Activity sampling sheet.

Activity sampling is carried out using the following steps:

1.  Define complex work cycles, such as days or weeks.
2.  Classify activities into working and non-working activities.
3.  Observe and code each activity.
4.  Summarise each activity and calculate time spent on each activity.

## Synthesis

*Synthesis is an indirect work measurement technique which uses established data in order to build up a standard time for an activity.*

The activity is divided into elements and can be further broken down into *motions. Predetermined time standards* (PTS) for particular elements or motions can then be used to reconstruct the original activity. Predetermined time standards are obtained by using *predetermined motion time studies* (PMTS) to time each motion.

A standard time can then be arrived at for an activity, without the need to directly time the activity in the field.

This technique can be useful when building up the time for a new activity. The concept behind the technique is that all activities are composed of a limited number of elements or motions, just like all music is composed of a limited number of notes.

## Analytical estimating

*Analytical estimating is an indirect work measurement technique where the standard times to carry out elements of an activity at a defined level of performance are estimated, partly from experience and practical knowledge, and partly from synthetic data.*

The accuracy of this technique will depend upon the experience and practical knowledge of the person carrying out the study because it relies on their judgement about the time needed for an activity.

This technique is obviously less accurate than the other techniques, but it may be the most appropriate one to use in some situations; for example, if it is a one-off activity or if there is not sufficient time to carry out a full study.

## Summary

Remember, work study is a complete topic in its own right and only the more common techniques have been examined in this chapter. However, by knowing the principles of work study, processes in the construction industry can be looked at with a more critical eye. The construction manager can then find ways to carry out work which are more efficient, more cost-effective, and safer.

---

### Keywords for work study

The following is a list of some keywords used in this chapter. Use the list to test your knowledge and, if necessary, consult the text to learn about the terms.

| | | |
|---|---|---|
| ASME | Install | Rating |
| Basic time | Multiple activity chart | Record |
| Efficiency | Process | Sampling |
| Examine | Productivity | Synthesis |

# 5 Sources of Finance

Construction activities are expensive and require considerable initial expenditure and financing during the construction process. Many of the parties involved in a construction project may require finance, including the building owner, the contractor and subcontractors.

Finance for construction can be described as the funds needed and the process of obtaining these funds, usually by borrowing. The majority of business organisations borrow money from some other organisation on the understanding that it must be repaid together with a charge for the service.

## Need for finance

It may be necessary for the building owner to borrow the finance on a long or medium term for the development and await a return on the investment before the finance can be repaid.

In the case of the contractors, they may have to use short-term finance such as overdrafts to finance the project before receiving payment from the building owner.

The principal reasons for seeking finance can be considered under the following headings:

- Insufficient internal finance
- Committed internal finance
- Competitive cost of external finance.

### Insufficient internal finance

Many organisations do not have sufficient funds in order to finance a construction activity. Building owners may require a building to use it themselves, or they may sell or lease the

68

building on completion. In each case the building owner will require considerable initial expenditure.

Contractors will require finance to fund the construction activity before they receive payment from the building owner.

## Committed internal finance

An organisation may have sufficient funds, but such funds may already be committed to other investments. If the funds of the organisation were used for construction, this might deplete the financial resources of the organisation and affect alternative investments.

## Competitive cost of external finance

It may be more profitable for building owners or contractors to borrow money from external sources rather than use their internal funds. The borrowing will allow an organisation to continue with other company investments which are yielding good returns.

This economic concept is termed *capital gearing*. It is normally better to be reasonably highly geared: to make use of other people's money rather than use one's own money, even though it may be expensive to do so.

---

### Some financial terms

#### *Shares*

A title of ownership in an organisation which gives its owner a legal right to part of the profits, usually in the form of dividends.

#### *Share capital*

The part of the capital of an organisation that comes from the issue of shares.

#### *Preference shares*

A share in an organisation which yields a fixed rate of interest rather than a varied dividend.

*(continued)*

> **Some financial terms** (*continued*)
>
> ### Dividend
>
> The distribution of part of the earnings of an organisation to its shareholders. The dividend is normally expressed as an amount per share.
>
> ### Capital gearing
>
> The ratio of the amount of fixed interest loan and preference shares in an organisation to its ordinary share capital.
> *Low gearing* is by a higher proportion of ordinary share capital.
> *High gearing* is by a higher proportion of fixed interest capital.

# Reasons for investment

**Reasons for investment**

*Cost of finance*
Rates of interest
Low risk investment
*Profit*
Maximise revenue
*Favourable economic climate*
Low cost of finance
Favourable conditions to
    borrow
Central government policies
Consumer demand exists
*Non-economic considerations*
Social
Efficiency
Economic
Expansion
Political
Marketing

Investment represents a form of saving. Prior to investing in construction activities, any organisation will need to consider the following factors:

- Cost of finance or rate of interest
- Profitability of investment or financial return
- Favourable climate of business confidence
- Non-economic considerations.

Some details of these factors are listed in the margin.

## Cost of finance or rate of interest

The cost of finance, which depends on the rate of interest, must be competitive in order for an organisation to invest. At times of low interest rates, organisations are attracted to borrow and invest in building stock. In general, large corporations and low risk activities enjoy greater access to funds and generally obtain such funds at a relatively low cost.

## Profitability of investment

The aim of all commercial organisations is to make a profit or financial return. Organisations such as building owners, contractors and financial institutions all expect to make a satisfactory return on their investment.

## Favourable climate of business confidence

A favourable economic climate will ensure confidence in the financial sector, which may have the following positive effects:

- Increase demand for property
- Low cost of finance
- Favourable conditions of borrowing.

Central government can encourage investment by adopting economic policies such as:

- Provision of grants for development
- Government investment in infrastructure projects
- Designating enterprise zones
- Provision of taxation subsidies.

## Non-economic considerations

Other factors may influence the need for investment, namely:

- Social considerations such as the need for hospitals, schools and road projects
- Economic considerations such as investment in infrastructure projects
- Political considerations such as favourable economic policies
- Improving efficiency
- Expanding existing business activities
- Launching a new service.

# Impact of finance

Each source of finance affects a business in the following three distinct ways:

- Risk
- Income
- Control.

## Risk

**Impact of finance**

*Risk*:
    minimise
*Income*:
    maximise
*Control*:
    monitor
    maintain
    report

The major risk is that the business will not be capable of meeting the financial commitments relating to the principal or interest. The total risk of the mix of capital sources is known as the financial risk of the business.

Financial risk can be explained as the capital invested in a project for which there is a substantial element of risk, especially with money invested in a new venture which may or may not provide an adequate financial return.

## Income

The income effect of financial sources relates to the cost of funds. Each source of finance has a cost attached to it. Reducing the cost of finance should increase the income of the owners.

## Control

The control effect of financial sources refers to new sources of finance affecting management or ownership control. It is the job of the financial manager to provide the required finance at minimum risk and minimum loss of control, while at the same time maximising the income to the owners.

# Finance for business

When funds are introduced into a business, whether or not in the form of cash, the firm can use those funds to acquire assets. *Share capital* raised and *long-term loans* will flow in as cash, whereas *trade credit* will allow further cash to be spent over a shorter term.

Amounts owed to the tax authorities can be used in the business until the due date of payment. The sale of assets will also provide cash to be used to acquire other *assets*, or to pay for *liabilities* that the organisation may have.

## Financial terms

### *Asset*

Any object that is of value to an organisation. In most cases assets are cash or can be turned into cash.

### *Liabilities*

These are normally taken to be a financial state of debt to another organisation. Financial liabilities can be grouped in the following four main categories:

- Current liabilities
- Deferred liabilities
- Contingent liabilities
- Secured liabilities.

### *Current liability*

An amount owed to the creditors of an organisation which is due to be paid within two months.

### *Deferred liability*

A prospective liability that will only become a definite liability if some future event occurs.

### *Contingent liability*

A liability that can be anticipated to arise if a particular event occurs; for example, a pending court case with an uncertain outcome.

### *Secured liability*

A debt against which the borrower has provided sufficient assets as security to safeguard the lender in the event of non-payment.

**Tangible assets include:**
Land and buildings
Plant and machinery
Trading stock
Investments
Debtors
Cash

**Intangible assets include:**
Goodwill
Patents
Trademarks

Organisations may be funded either *internally* or *externally*.

## Internal finance

Internal funding is finance provided by the operations of the organisation itself. The cost of internal finance is sometimes referred to as *opportunity cost*. Opportunity cost can be described as the value of the alternative that has been sacrificed in order to pursue a certain course of action.

### Internal short-term funding

Where a business has been operating successfully, the owners may have decided not to pay all of the profits out, but to reinvest in the business. These retained profits form the major source of internal finance to be used for future expansion.

Short-term funding also generates credit for the contractor and building owner respectively thus allowing the surplus finance to be used in the short term. Examples of such short-term finance include:

- Paying for materials in arrears
- Receiving stage payments in arrears.

### Internal medium-term funding

Internal medium-term funding is a less common source. The principal form of medium-term finance is 'inter-company loans' within an organisation or large group.

### Internal long-term funding

Internal long-term finance is most often provided by 'retained earnings'. Retained earnings are profits after tax which are kept within an organisation. These profits can be an important source of funds for expansion and take-over activities.

It is no easier to obtain internal finance than external finance and in some instances more difficult. There may be some willingness to be sympathetic to riskier projects, but the rate of return expected from internally financed projects is usually substantially higher than from externally financed projects.

## External funding

External funding is finance provided by sources outside the organisation, such as banks or building societies.

The use of external funding is determined by other factors such as the following:

- Corporation type, for example partnership or public limited company
- Purpose of funding
- Risk associated with funding.

Types of finance

| Term of finance | Type of finance | Purpose of finance |
| --- | --- | --- |
| Short term | Short-term loan<br>Factoring<br>Overdraft | Purchase of machinery<br>Raising finance against debtors<br>Short-term cash deficit |
| Medium term | Hire purchase<br><br>Leasing<br>Loan finance | Purchase of company vehicles and plant<br>Purchase of plant and machinery<br>Improvement to premises |
| Long term | Mortgage<br>Venture capital | New business premises<br>New machinery, investment in new technology |

# Terms of finance

A contractor may require finance for a variety of projects over different periods of time; this may be grouped into three types:

- Short-term finance
- Medium-term finance
- Long-term finance.

## Short-term finance

Finance may be required for up to one year for a short term only, to pay for land and building work from inception to eventual disposal. This capital is normally raised to cover exceptional demand for funds over a short period. This form of finance is the most expensive and is usually used to assist a company with its cashflow and working capital requirements.

**Short-term finance**
Trade credit
Bank overdrafts
Factoring
Discounting
Taxation
VAT
Bank loans
Accrued expenses
Employee wages and salaries

## Types of short-term finance

### Trade credit

Value of goods that a trader or organisation allows a customer before requiring payment.

### Bank overdrafts

A loan made to a customer with a cheque account in which the account is allowed to go into debt, usually to a specified limit.

### Factoring

The sale of trade debts at a discount to an agency which assumes the task of recovering the debt and accepting the credit risk. This process gives the original supplier quick repayment of a guaranteed amount.

### Discounting

A reduction in the price of goods for buyers who pay by cash (cash discount), for members of the trade (trade discount), or for buying in bulk (bulk discount). These processes encourage early repayments and larger purchases.

### Taxation

A levy on individuals of corporate bodies made by central or local government. Taxation is normally paid in arrears and the sums owed can act as a source of short-term finance.

### Value Added Taxation

VAT is a tax levy added to goods or services at each step in the chain of original purchase, manufacture, and final sale or service. Each trader in the chain collects the tax and has use of the money before it must be paid to the Customs and Excise.

### Bank loan

Money lent on condition that it is repaid, either in instalments or all at once, on agreed dates and usually on condition that the borrower pays the lender an agreed rate of interest.

*(continued)*

---

**Types of short-term finance** (*continued*)

*Accrued expenses*

Regular expense payments owed by an employer to an employee, but usually paid monthly in arrears.

*Employee wages and salaries*

Regular payments made by an employer to an employee under a contract of employment. These are usually paid monthly in arrears.

---

## Medium-term finance

Medium term normally refers to a time period of one to five years which is required to finance specific expansion projects. This period is normally too long a time to be covered by term funding, but insufficient to warrant the use of shares. Examples of medium-term finance are listed in the margin.

**Medium-term finance**
Hire purchase
Leasing
Medium-term loans
Debentures

---

**Types of medium-term finance**

*Hire purchase*

A method of buying goods, such as vehicles, in which the purchaser pays a deposit and takes immediate possession of the goods. The purchase is completed by a series of regular instalments.

*Leasing*

The hire of equipment, such as cars and machinery, to avoid the capital cost of ownership.

*Debentures*

A common type of fixed-term loan made to a company. The debenture bond document usually specifies a fixed rate of interest which is paid before any dividends are paid to shareholders.

## Long-term finance

Long term refers to a time period of more than five years, and is generally used for long-term investment purposes. The traditional method of providing long-term finance is by means of a mortgage. Insurance companies and pension funds provide most mortgages in the commercial and industrial sectors, while building societies provide the largest share of mortgages for the private housing building market.

---

**Types of long-term finance**

*Mortgage*

A loan which uses the value of a property as security for the repayment. The loan is usually used to purchase the property.

*Sale and leaseback*

Finance released by an organisation selling land or buildings to an investor, such as an insurance company, on condition that the investor leases the property back to the organisation under agreed terms.

*Capital grants*

Finance made available by central or local government to encourage specific initiatives or investments, such as building a factory.

*Preferential share capital*

That part of the share capital of an organisation which is held as *preference shares* which have priority over ordinary shares.

*Equity share capital*

That part of the share capital of an organisation which is held as *ordinary shares*.

*Share issues*

Finance is raised when a limited liability company sells shares to the public.

*(continued)*

**Types of long-term finance** (*continued*)

*Bonds*

A loan backed by a certificate in which the borrower promises to repay on a certain date. Bonds issued by governments, local authorities and companies usually pay fixed rates of interest.

*Pension funds*

State and private pension contributions which are invested to give as high a return as possible to provide the funds for payment of pensions.

*Venture capital*

Capital exchanged for shares in a project in which there is a substantial element of risk, such as with a new type of business or with an expanding business.

*Government incentives*

Finance made available by international organisations, central government or local government to encourage investments in specific initiatives.

# Sources of finance

Financial institutions, such as banks and building societies, advertise widely in the press and on television about the availability of funds for borrowers. However, one should not be misled by the advertisements into thinking that finance is easily obtained. To 'source' finance can be difficult for the borrowing organisations. Although the 'sources' will lend money, they will do so only if they believe that there is a very high probability that the loan will be repaid.

The principal origins of finance include:

- Financial institutions
- Central government
- Local government.

# Financial institutions

*A **financial institution** is any organisation, such as a bank, building society or finance house, that collects funds from individuals, other organisations, or government agencies, and invests these funds or lends them on to borrowers.*

Some financial institutions, such as brokers and life insurance companies, are non-deposit-taking. They fund their activities and derive their income by selling securities or insurance policies, or by undertaking brokerage services.

At one time there was a clear distinction between deposit-taking and non-deposit-taking financial institutions and they were kept separate by regulations. This is no longer the case; brokers and other companies now often invest funds for their clients with banks and in the money markets. Example of such institutions are listed.

---

**Financial institutions**

*Bank*

A commercial institution licensed to take deposits. Activities include making and receiving payments on behalf of customers, accepting depots, and making short-term loans to individuals, companies and other organisations.

*Building society*

A financial institution traditionally intended to provide long-term mortgages for house buyers and home improvers. Funds are raised by savings schemes.

*Finance house*

An organisation that mainly provides hire-purchase and leasing agreements. Many finance houses are owned by commercial banks.

*Merchant bank*

A bank that originally specialised in financing foreign trade. They have now diversified into hire-purchase finance long-term loans, venture capital, advice to companies about flotations on the stock market, underwriting new issues of shares, management of investment portfolios and unit trusts.

*(continued)*

---

---

**Financial institutions** (*continued*)

*Pension funds*

Funds managed by individual organisations working closely with insurance companies and investment trusts. Such funds have a large influence on many of the securities traded on the Stock Exchange.

*Special financial institutions*

Organisations such as the Industrial and Commercial Finance Corporation (ICFC) provide financial assistance for small and medium-sized companies in all industries.

---

## Central government

The government is a large client of the construction industry. Any changes in its policy towards building and engineering projects are likely to have a considerable effect on the performance of the industry. For example, a reduction of government investment in housing, roads, schools, colleges, health care or other schemes will have a direct impact on the availability of finance incentives for organisations wishing to invest in such projects. Examples of government incentives are listed in the margin.

**Government incentives**
Tax incentives
Regional development funds
Grants
Subsidies

## Local authorities

Local government policy which is implemented by local authorities, such as town councils, may also encourage local investment and offer funds to encourage local investment. Examples of local authority incentives or schemes are listed in the margin.

**Local incentives**
Inner city development
Grants
Long-term loans
Construction of civic
  buildings
Enterprise boards
Local enterprise agencies
Economic development
  corporations

## Other sources of finance

Examples of other sources or origins of finance include:

- Hire purchase
- Factoring
- Venture capital
- Issue of shares.

# Conditions attached to finance

Before an organisation approaches a financial institution, the prospective borrower will need to consider the following questions:

- Why is the finance required?
- How much finance is required?
- How and when is the money to be repaid?
- Will the organisation require further borrowing during the loan period?
- What security can be offered to the financial institution?

The finance agreement will normally take the form of a legal contract between the financial institution and the borrower, and this agreement may need a review or renewal every twelve months.

### Information from borrower

Typical information required by a financial institution prior to providing finance is listed below:

- Copy of audited accounts of last three years of trading
- Financial accounts
- Forecasts of anticipated profits on future contracts
- Cashflow forecast
- Schedule of work in progress
- Schedule of fixed assets and equipment
- Schedule of borrowings
- Capital expenditure plans
- Report on company personnel and structure
- Copy of strategic plan and organisational objectives.

### Typical conditions attached to borrowing

Some of the more typical conditions attached to finance include the following:

- Repayment
- Interest
- Breach of agreement
- Varying agreement
- Applicable law

- Security
- Protection insurance
- Commission
- Early settlement
- Taxation
- Notice
- Use of finance
- Eventual ownership.

## Summary

The aim of financial sourcing is to ensure that finance is available with the following features:

- Sufficient for planned projects
- Appropriate types and mixes of finance
- Finance available as and when required
- Controls available to ensure effectiveness.

The aim is to use the finance for the most efficient operation of the organisation.

---

**Keywords for sources of finance**

The following is a list of some keywords used in this chapter. Use the list to test your knowledge and, if necessary, consult the text to learn about the terms.

| Capital gearing | Assets | Shares |
| Share capital | Liabilities | Leasing |
| Dividend | Factoring | Trade credit |
| Risk | Debentures | |

---

# 6 *Financial Management*

Construction often involves complex and intensive site operations linked to large sums of money which need careful financial management. There is a great variety of patterns of construction work and finance which lead to a wide selection of commissioning options available to the client.

This chapter examines the principal management techniques used in controlling finance during the overall construction process and the principal procurement options available to the client.

*Financial management can be described as the management of the money of an organisation in order to achieve the financial objectives of that organisation.*

The objectives of an organisation can be divided into two main categories:

- Financial objectives
- Non-financial objectives.

Examples of the principal financial and non-financial objectives are listed in the margin.

**Financial objectives**
Maximum profit
Interest of shareholders

**Non-financial objectives**
Welfare of employees
Welfare of management
Welfare of society
Provision of a service
Responsibility to customers
   and suppliers

## Management processes

The process of managing finance involves two principal aspects:

- Financial planning
- Financial control.

## Financial planning

Financial planning involves forecasting the future needs of the organisation. The process is linked to the objectives of management which in turn will formulate the policy of the organisation. Typical issues of financial management include the following:

- The amount of finance required
- When the finance is required
- How the finance is raised
- What type of finance is available
- How the finance is allocated.

Good financial planning ensures that sufficient funding is available at the right time to meet the needs of the client or contractor for short, medium and long-term capital. Short-term finance may need to be made available for the purchase of plant or machinery. In the medium or long-term the organisation may need to finance the construction of a new head office.

Planning the flow of cash in the organisation is one of the major functions of management. Incorrect cashflow can cause loss of profit and the subsequent failure of a company.

## Cashflow forecasting

A cashflow forecast is a statement of the estimated cashflows in and out of an organisation. The forecast can give the organisation an early indication of any shortages or surplus of cash.

The cash flowing in, known as *positive cashflow*, is the cash received. Cash flowing out, known as *negative cashflow*, is the cash paid out. The difference between cash flowing in and out is known as the *net cashflow*. This may be a surplus (more cash in than out), or a deficit (more cash out than in). These terms should not be confused with 'profit' and 'loss'.

### *Preparation of cashflows*

The process of preparing a cashflow involves predicting cashflows for individual projects which, in turn, are amalgamated into the master cashflow for the company. The master cashflow will also provide for those general overheads which cannot be allocated for individual projects. In a very large company a separate cashflow for overheads may be needed.

To prepare a cashflow forecast the following type of information is needed:

- Contract budget in monthly or cumulative form
- Contract period

**Cash inflows and outflows**

*Cash inflows*
Valuations of works in
  progress
Stage payments
Release of retention
  monies
Final account payments
Loans for other financial
  institutions
VAT refunds

*Cash outflows*
Salaries
Payments to subcontractors
Payments to suppliers
Insurance premiums
Overhead costs
Loan repayments

- Payment periods
- Retention details
- Defects liability period
- Anticipated profit release
- Delay in meeting the cost commitment.

---

### Explanation of terms

*Retention*

A percentage of each interim payment due to a contractor which the client is allowed to hold back under the terms of the construction contract. The contractor is normally entitled to the release of one half of the retention monies when the works are completed and the release of the balance at the end of the defects liability period.

*Defects liability period*

A period, normally six or twelve months, after the completion of the works when the contractor must make good any defects in the building for which he is liable. Once the defects have been made good to the satisfaction of the architect, a certificate of making good defects will be issued and the remaining retention monies held will be released to the contractor.

---

### Use of cashflow

The cashflow forecast will enable an organisation to carry out the following functions:

- Ensure that cash is available to meet day-to-day requirements
- Meet periodic demands for longer term cash requirements
- Plan payments for larger financial commitments
- Repayment of debts
- Transfer surplus cash for investment
- Ensure temporary credit arrangements are in place when required.

An example of a cashflow forecast is shown in Figure 6.1. Effective cash planning will give visible benefits to the business such as increased efficiency and a more stable financial structure.

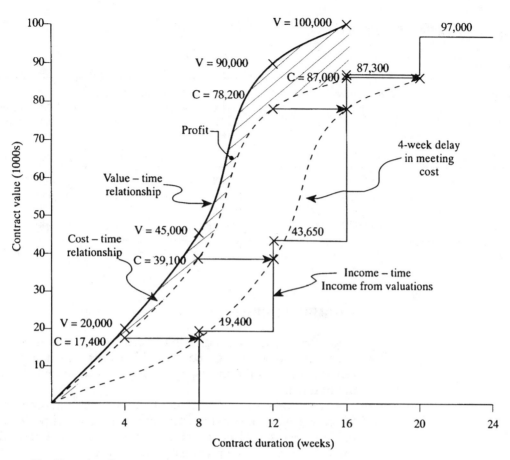

*Key:* V = value, C = cost

**Figure 6.1** Cashflow diagram.

## Example of cashflow

### Details of case study

| | |
|---|---|
| Contract value | £100,000.00 |
| Contract period | 16 weeks |
| Payment intervals | monthly |
| Retention rate | 3% |
| Profit allowances in estimate | 15% |
| Delay in meeting cost (average payment delay) | 4 weeks |
| Interest payable in financial contract | 20% |

(*continued*)

**Example of cashflow** (*continued*)

*Cashflow table*

| Valuation | Cumulative value | Cumulative cost | |
|:---:|:---:|:---:|:---:|
| | | *(cumulative value ×100/100+ profit)* | *(Cumulative value less 3% retention)* |
| *(£)* | *(£)* | *(£)* | *(£)* |
| 1 | 20,000 | 17,400 | 19,400 |
| 2 | 45,000 | 39,100 | 43,650 |
| 3 | 90,000 | 78,200 | 87,300 |
| 4 | 100,000 | 87,000 | 97,000 |

# Financial control

Once the necessary funding has been raised, it is necessary to keep the finance under control. Measures which ensure that finance is put to the most profitable use need to be devised and maintained.

The process of financial control involves devising systems which regulate financial matters and check actual performance. An important tool is the comparison of actual financial performance against the planned performance.

Part of the system of financial control in an organisation is a budgetary control system which measures the progress of the initial financial plan.

## Budgeting

**Resource budgets**
Labour
Materials
Plant
Site overheads
Cash

Budgeting can be explained as the forecasting and monitoring of both the income and expenditure of the organisation.

*Budgeting is the process of financial control whereby the actual income and expenditure for a period is compared with an appropriate budget allowance for each item in the same period.*

The budget is normally prepared prior to the start of a trading or operating period and sets out the objectives, activities and policies to be carried out during that period of the business. Annual budgets for the whole year are commonly broken down into shorter control periods of operation such as months or, in some cases, weeks.

It is important for the contractor to know the amount of capital that is needed for a project and when it is required. In order to find these figures, the contractor needs to draw up a programme of work and use this programme to find the rate of expenditure and rate of income over a particular time period.

The difference between the expenditure and the income will give the amount of capital required. Separate budgets can be established for a number of resource headings, such as those listed.

The main aims of budgetary control are identified below:

- To ensure that planned profits are maintained
- To provide future cost information
- To provide a continuous comparison of actual cost to planned cost
- To enable corrective action to be taken.

Good financial control will involve the setting up of a cost control system that suits the particular needs of an organisation. In order to regularise the costing system the organisation needs to adopt a number of cost headings against which to record costs (see Table 6.1).

**Table 6.1**   Typical cost centres

| Cost code number | Description | Subdivision |
|---|---|---|
| 10 | Labour | |
| 20 | Materials | Directly employed<br>Labour only subcontractors<br>Labour overheads |
| 30 | Plant | Mechanical<br>Non-mechanical maintenance |
| 40 | Subcontractors | Domestic subcontractors<br>Nominated subcontractors |
| 50 | Salaries | Head office salaries<br>Site-based salaries<br>Hourly paid wages<br>Directors' salaries<br>Fees to consultants |
| 60 | Assets | |
| 70 | Staff training | |
| 80 | Expenses | |

# Construction project finance

Financing the construction process involves the use of two fundamental principles of construction costing:

- Cost planning
- Cost control.

Cost planning and cost control are normally carried out by the Quantity Surveyor and the processes need to be effective for the following reasons:

- Client demands
- Current economic conditions
- Increased competition
- Need for control of construction costs
- Distribution of costs in a balanced manner.

The process of cost planning and cost control can be conveniently represented by using the Royal Institute of British Architects Plan of Work for the design team. The stages of cost planning and cost control and their relationship to the RIBA plan are shown in Table 6.2.

## Design stage finance

### Inception

Inception is the first stage in the design sequence, where the client approaches the architect with a list or brief of what is required. The client needs an indication of the total cost at this stage.

**Preliminary cost estimate breakdown**

Basic estimate
Adjustments
   Market conditions
   Size, number of storeys
   Specification changes
   Inclusions
   Exclusions
   Building services charges
   Site conditions
   Pricing level
Professional fees

### Feasibility

At this stage the client will need assurances that the project under consideration is feasible. The Quantity Surveyor is normally expected to prepare a feasibility report.

### Outline proposals

At this stage of the design, the level of information available to the design team and Quantity Surveyor should be sufficient to enable the lump sum estimate previously prepared to be allocated to cost headings. A convenient cost breakdown would be to use the eight major groups of elements as devised by the Building Cost Information Service (BCIS), which is a service provided by the Royal Institution of Chartered Surveyors (RICS). This elemental breakdown provides a useful cost framework for the Quantity Surveyor to monitor and control cost as the design develops.

**Feasibility stage cost estimate**

Office accommodation
External works
Preliminaries
Contingencies
Allowances
   Market
   Size, number of storeys

**Table 6.2** Stages of cost planning and control

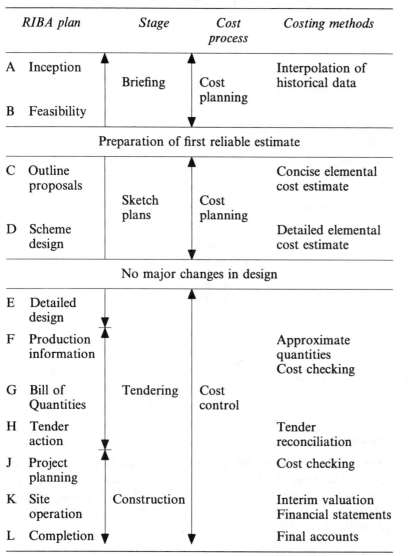

| RIBA plan | | Stage | Cost process | Costing methods |
|---|---|---|---|---|
| A | Inception | Briefing | Cost planning | Interpolation of historical data |
| B | Feasibility | | | |

Preparation of first reliable estimate

| C | Outline proposals | Sketch plans | Cost planning | Concise elemental cost estimate |
|---|---|---|---|---|
| D | Scheme design | | | Detailed elemental cost estimate |

No major changes in design

| E | Detailed design | | | |
|---|---|---|---|---|
| F | Production information | | | Approximate quantities Cost checking |
| G | Bill of Quantities | Tendering | Cost control | |
| H | Tender action | | | Tender reconciliation |
| J | Project planning | | | Cost checking |
| K | Site operation | Construction | | Interim valuation Financial statements |
| L | Completion | | | Final accounts |

**Feasibility stage cost estimate** (*continued*)
Specification changes
Inclusions
Exclusions
Building services charges
Site conditions
Price risk
Design risk

**Elemental cost estimate breakdown (BCIS elements)**
Substructure
Superstructure
Internal finishes
Fittings and furniture
Services
External works

**Elemental cost plan breakdown**
Substructure
Superstructure
  Frame
  Upper floors
  Roof
  Stairs
  External walls
  Windows and external doors
  Internal walls and partitions
  Internal doors
Internal finishes
  Wall finishes
  Floor finishes
  Ceiling finishes
Fittings and furnishings
Services
External works
Preliminaries
Contingencies

### Scheme design

As the design information develops, the Quantity Surveyor can break down the cost into a greater number of cost centres in order to achieve control on the distribution of the cost. (A useful framework is to use the BCIS standard list of elements as listed nearby.) The object of splitting the cost down into so may divisions is to ensure a better-balanced design and to ensure that the client gets value for money.

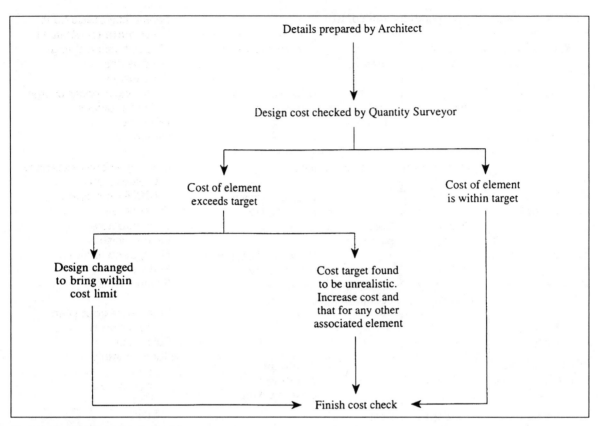

**Figure 6.2**   Outline of cost checking process.

### Detailed design

During the detailed design the Quantity Surveyor will be involved in cost checking the monies allocated to the various elements to ensure that the sums allocated are still realistic (see Figure 6.2). The Quantity Surveyor may use approximate quantities to check the individual costs.

## Tendering stage finance

The tendering stage concerns the period from the production of the tendering documentation to the selection of the successful contractor.

### Production information

After the final cost check, the Architect and the other members of the design team start to prepare the detailed information necessary for the Quantity Surveyor to prepare the Bill of Quantities. The typical documentation produced includes:

- Tender drawings
- Specification.

## Bill of Quantities

The Bill of Quantities is prepared, checked, printed and sent out to the tenderers. Further cost checks are carried out, as the completed work sections are measured, and costed and compared with the budgetary allowances.

## Tender action

The Quantity Surveyor compares the submitted prices with the cost plan. On receipt of the tender the Quantity Surveyor evaluates the tenders and, after checking, recommends to the client a suitable contractor to agree the contract and to commence the works.

# Construction stage finance

## Site operation

Once the work commences on site the Quantity Surveyor needs to control the cost of the project. This is done by compiling interim valuations and financial statements which are produced at regular intervals. The financial statements, together with the cashflow forecast, show the estimated final cost of the project, taking into account any variation orders that may have been issued, together with adjustments for prime cost and provisional sums.

## Completion

The Quantity Surveyor prepares a *final account* which includes the payment of the contract work, changes to the contracted works and any *claims* for delays or disruption to the works.

---

### Cost control terms explained

#### Interim valuations

An interim valuation of the work, usually made monthly, prepared by the consultant Quantity Surveyor on behalf of the main contractor. A payment is recommended by the Quantity Surveyor to the Architect who, in turn, issues an 'Interim Certificate' which is then forwarded to the client as authorisation to make payment.

*(continued)*

**Cost control terms explained** (*continued*)

*Financial statements*

A financial report normally prepared by the consultant Quantity Surveyor to update the client on the current and projected expenditure on the construction project.

*Variations*

Authorised changes to the contracted works confirmed in writing by an Architect's instruction. It is normally the responsibility of the consultant Quantity Surveyor to value and agree the value of these changes with the contractor.

*Final account*

A final claim for payment of work which is normally prepared by a consultant Quantity Surveyor on behalf of the contractor. The account should include all claims in respect of all work carried out by the main contractor and any specialist subcontractors or suppliers.

*Claims*

A claim made by the contractor for additional payment, allowed for in the contract, because of unexpected events such as the issue of variation orders and disturbances to the regular progress of the work.

# Procurement options

Procurement can be explained as the combination of activities undertaken by a client in order to get a building constructed. The term is more directly defined as follows:

**Procurement** *refers to the choice of contractual arrangements available for selecting a contractor to construct a building.*

> ## Procurement systems
>
> ### *Traditional*
>
> The client appoints an Architect or other professional to produce the design, then to select the contractor and supervise the works until completion. The contractor generally has no responsibility for design and is selected on a competitive basis.
>
> ### *Design and build*
>
> The client appoints a single organisation to carry out both the design and the construction under a single contract.
>
> ### *Management*
>
> Because of the size and complexity of the project, the client appoints a management consultant to take over the managerial role of the construction project. The contractor works alongside the cost consultants, providing a construction management service. The managing contractor does not undertake either the design or direct construction work.

Common systems of procurement are described below and various combinations and adaptations of these systems are also in use.

In making a choice of procurement system, the client needs to consider and balance three important factors:

- Cost
- Time
- Quality.

The importance of a particular factor for a particular client will probably suggest where design responsibilities are to rest. This priority indicates the most suitable procurement system and leads to the selection of a suitable form of contract.

Many other factors, such as the ones listed, should be considered before the procurement system is selected.

**Procurement factors**
Cost – Time – Quality
Type of client
Size of project
Complexity
Flexibility
Competition
Risk
Design responsibility
Market conditions

## Traditional procurement

In a traditional procurement arrangement, a client appoints a design team to produce the production information needed for the project. *Tendering* involves selected contractors offering a price for which they will carry out the work described, in accordance with the conditions of the contract.

A tender is chosen which represents the best overall value for money and a contractor is appointed to proceed with the building project.

### Advantages

- High degree of certainty on cost and time before commitment to build
- Clear accountability and control
- Competitive pricing
- Combination of best consultant and contracting skills
- Flexibility in design development.

### Disadvantages

- Greater coordination and control required
- Little opportunity for contractor to contribute at design stage
- Relatively slow system.

## Design and build

In a design and build arrangement one contractor accepts the responsibility for both the design and construction of a building in order to meet the requirements of the employer.

### Advantages

- Early certainty of overall price
- Single responsibility
- Early focus on needs of client
- Speed in design and construction
- Overlapping of design and construction
- Simpler coordination
- Economy in design
- Buildability (ease of construction)
- Encourages cooperation within the team.

### Disadvantages

- Difficulty in evaluating tenders
- Variations might be expensive
- Possible reduction of quality standards
- High cost of tendering for contractors
- High risk for contractor in pricing.

## Management

A management arrangement is one where there is a managerial approach to construction. A management specialist is appointed early in the project to help ensure *buildability* or ease of construction. The purpose of the specialist is to perform a leadership function and to work in close liaison with the client's team of professionals.

### Advantages

- Enables early involvement of management contractor
- Early start on site possible
- Possibility of overlapping the design and construction
- Good cost control procedures
- Closer relationships between design and construction elements
- Useful in complex construction projects.

### Disadvantages

- No fixed price agreed
- Construction starts with incomplete design
- Employer committed to start building on cost plan, project drawings and specifications
- Not recommended for smaller projects
- High risk for client.

# Contract options

Once a suitable procurement system is selected, the client must decide on a 'type of contract' and a 'form of contract'. The contract 'type' stipulates the provisions under which the contractor and the client agree to enter under a contract of agreement. The essential question considered in selecting a type of contract is how the client intends to pay the contractor.

The client has the following broad choices of contract arrangements.

### Lump sum

Payment by means of a lump sum which is determined before the contract starts.

### Measurement

Payment by means of remeasuring the works as they progress by means of an agreed schedule of rates.

### Cost reimbursement

Payments arrived at on the basis of actual costs of labour, plant and materials, to which a fee to cover overheads and profit is added.

Once the procurement system and the type of contract have been selected it is necessary to select an appropriate 'form' of contract (see Table 6.3). The client may decide to use one of the many standard forms of building contract, or to use an in-house contract or a non-standard form.

**Table 6.3**   Some choices of contractual options

| Procurement method | Type of contract | Forms available |
|---|---|---|
| Traditional | Lump sum | JCT1980 (with Quantities) JCT1980 (without Quantities) IFC 1984 |
|  | Measurement | JCT 1980 (Approximate Quantities) |
|  | Cost reimbursement | JCT Fixed Fee Prime Cost 1976 |
| Design & Build | Lump sum | JCT with Contractor Design 1981 |
| Management contracting | Management fee | JCT Standard for Management Contract 1987 |

Note: JCT = Joint Contracts Tribunal; IFC = Intermediate Form of Contract.

---

### Keywords for financial management

The following is a list of some keywords used in this chapter. Use the list to test your knowledge and, if necessary, consult the text to learn about the terms.

| | | |
|---|---|---|
| Cashflow forecasting | Budgeting | RIBA Plan of Work |
| Retention | Cost headings | Cost planning |
| Defect liability period | Final accounts | Variation orders |
| Interim valuations | Lump sum contract | Cost plus contract |

# 7  Cost Planning

Planning the cost of construction projects concerns keeping the cost within a predetermined cost estimate during the design stages. This planning normally involves the preparation of a cost plan and the carrying out of cost checks as explained in the previous chapter.

The process of cost planning involves the preparation of an *approximate estimate*. The object of the approximate estimate is to provide an estimate of the probable cost of a project before the design has commenced.

Cost planning has three main functions:

- Early indication of cost prior to design development
- Confirmation of financial commitment for building owner
- An 'early warning' system for any changes needed in design or construction techniques.

When preparing a cost estimate the following guidelines must be followed:

- Specify date of preparation
- State source of cost information
- State period for which price is valid
- Identify any exclusions
- State assumed tender date
- Update estimate for time, quantity and quality.

The factors which normally influence buildings costs are listed.

**Building cost factors**
Taxation
Economy
Location
Market
Tendering
Specification
Construction type
Construction method
Design factors: shape, size, height, number of storeys
Site conditions
Contract type
Procurement type

## Accuracy of cost estimates

The accuracy of a cost estimate depends upon the level of information available at the time of the preparation of the

## Cost estimating techniques

Unit method
Cube method
Superficial method
Storey enclosure method
Approximate quantities
Elemental method
Comparative method
Interpolation method
Elemental cost analysis
Cost modelling

estimate. The accuracy of the estimate will improve as the project design develops.

A number of techniques are used for preparing cost estimates, some of which are more sophisticated than others.

During the design stage the surveyor is likely to use several different methods at different stages in the RIBA Plan of Work. An illustration of the role of approximate estimating techniques is shown in Figure 7.1.

The initial estimate provided at the feasibility stage of the project will generally be calculated using the *unit method* of approximate estimating. This method allocates a cost to each accommodation unit, such as a bed or a seat, and is based upon historic data from similar projects which has been updated to make allowance for time, quality and quantity.

At this stage of the design little information is available. This method of estimating can make little allowance for such

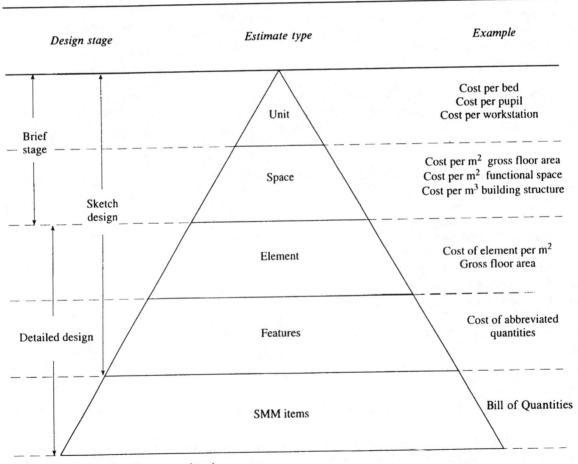

**Figure 7.1**  Traditional early stage estimating.

variables as size, shape, construction method and the time period before tender action is taken.

In order to make allowance for the above the estimator has to use experience and judgement to allow for risk. As the design develops, the level of information available increases quickly, and so the price forecast can be updated and will become more accurate.

Even after the initial outline proposals have been presented, the price forecast can be recalculated on a more accurate basis using *elemental cost planning*. The cost information used is based on historical cost data from a series of similar projects which have been updated for time, quality and quantity.

At the outline design stage, the project costs can be split into elements and from the historic data of completed projects each element has its cost expressed as a cost per m². These element rates can then be applied to the present project, again making an allowance for risk.

Once a particular construction method has been established and the scheme design begins, the estimator will switch to a more accurate form of estimating. When more formal drawings are produced, it is possible to measure the quantities and therefore produce an estimate based upon *approximate quantities*. These can be priced in line with current tenders before being adjusted for time and value of money. The risk allowance at this stage will be reduced as more details are provided by the designers.

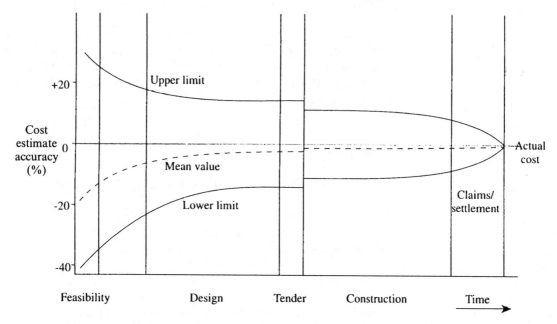

**Figure 7.2**  Accuracy of cost estimates.

As the design and specification are completed and accurate quantities are taken off, the price forecast can be obtained from competitive tenders; however, a smaller allowance should still be made for settlement of the final account.

Figure 7.2 shows the relationship between the accuracy of the price forecasting and the stage of the project. As more detailed design information is made available, the cost estimate becomes more accurate, as the figure shows.

# Adjustment of historical cost data

**Adjustment factors**
Price
Quality
Time

The historical data from an already-analysed building may be used to prepare cost estimates, but the proposed new building may differ in the following ways:

- Size
- Shape
- Number of storeys
- Specification.

It is also necessary to make adjustments for the three factors of price, quantity and time.

## Price

Historical costs reflect the market price level at the date of tender of the building analysed. It is necessary to update these costs by use of an appropriate index. It may also be necessary to adjust the price level in the second period for 'price and design risk'.

## Quantity

Quantity can be adjusted in the following three ways:

- By proportion
- By approximate quantities
- By inspection.

### Proportion

The method of proportion involves the use of *quantity factors* which represent a mathematical relationship between two measurement variables. Examples of such factors together with a typical proportional adjustment are shown below.

### *Approximate quantities*

This method involves taking off dimensions from drawings. At the early stage of design, however, the information is not normally detailed enough to use. We will see the use of approximate quantities at a later stage in the design.

### *Inspection*

It is usual to examine a number of analyses of similar projects to obtain a range of costs, and then to exercise personal judgement of the figure to use in the cost plan.

## Quality

Quality may also be adjusted by proportion, by approximate quantities or by inspection. However, as the specification is not normally available at the early design stages, it is more usual to make the quality adjustment by coupling inspection with simple calculations. An example of a quality adjustment is shown in the example.

---

### Examples of price adjustment

#### *Index adjustment*

Adjust using tender index   Adjust for price and design risk
    Add 8%                Add 5 %

$$\text{◄►◄} \longrightarrow$$

| Historic building tender date | Proposed tender date | Proposed completion date |

$£16,600,000.00 \times 1.08 = £17,928,000.00 \times 1.05$
$\qquad\qquad\qquad\qquad\quad = £18,824,400.00$

#### *Rate adjustment*

*Building A (historic building)*

Gross floor area of building $1310\,m^2$

                                                     *(continued)*

---

**Examples of price adjustment** (*continued*)

External wall area $760\,\text{m}^2$

$$\text{Quantity factor} = \frac{760}{1310} = 0.58$$

*Building B (proposed building)*

Gross floor area of building $1410\,\text{m}^2$
External wall area $420\,\text{m}^2$

$$\text{Quantity factor} = \frac{420}{1410} = 0.30$$

If the rate per $\text{m}^2$ of external wall for the historic building A was £25.50, then the external wall rate for the proposed building B is calculated as follows:

$$£25.50 \times \frac{0.30}{0.58} = £13.19/\text{m}^2 \text{ for the proposed building}$$

*Quality adjustment*

Suppose there is a quality change in the cost of facing bricks, the comparable material unit rates might be £4.00/$\text{m}^2$ for the historic building and £5.10/$\text{m}^2$ for the proposed building. The adjustment is calculated as follows:

$$\frac{£5.10}{£4.00} = 1.28, \text{ which is a 28\% increase}$$

*Overall adjustment*

Rate for historic building: £25.50 per $\text{m}^2$

Price factor: 1.08 (8% increase)

Quantity factors:
   Historic quality factor: 0.58
   New quality factor: 0.30

Quality factor: 1.28 (28% increase)

The rate for the proposed building is calculated as follows:

$$£25.50 \times 1.08 \times \frac{0.30}{0.58} \times 1.28 = £18.26 \text{ per } \text{m}^2$$

# Approximate estimating methods

## Unit method

The unit method of approximating lends itself to certain types of buildings where the building's function can be expressed in terms of units such as beds or seats. Examples of the technique's application are listed.

The unit method is useful early in the design stage where a client may only be able to express the requirements in simple unit terms.

### Advantages of unit cost method

- Provides a convenient form of stating a cost limit
- Provides a simple method of comparing different schemes
- Speed of application.

### Disadvantages of unit cost method

- Lack of precision
- Unsuitable for estimating the cost of individual and different buildings
- Difficult to adjust unit cost accurately to take account of variables
- Difficult to relate to changes in storey heights
- Difference in time excluded
- Location of the site excluded
- Method of construction excluded.

**Types of units**
Number of beds in a hospital
Number of seats in a cinema
Number of parking spaces in a car park
Number of bedrooms in a hotel
Number of pupil places in a school

---

### Example of unit cost method

Assume that a hospital project intended to cater for 200 patients has been completed for a total cost of £16,600,000. The unit cost of each hospital bed is:

$$\frac{\text{Building cost}}{\text{Number of beds provided}} \quad \frac{£16,600,000}{200} = £83,000 \text{ per bed}$$

---

## Superficial method

The superficial method is the most popular method of approximate estimating. The total floor area of a building is measured between the internal faces of the enclosing walls. This area is then multiplied by a calculated unit rate per square metre in order to obtain the probable cost.

### Advantages of superficial method

- Simple method
- Easiest to remember
- Rapid method
- Widely used by Architects and Quantity Surveyors
- Readily understood by the employer
- Wide availability of data.

### Disadvantage of superficial method

- Buildings must be of similar shape, size, height and specification.

---

**Example of superficial method estimate**

Assume that a hospital has been built with the following floor areas (measured within external walls):

Ground floor plan: $750 \, \text{m}^2$
First floor plan: $750 \, \text{m}^2$

So the total floor area is $1500 \, \text{m}^2$.

The cost of the hospital can now be expressed in terms of price per $\text{m}^2$ of floor area:

$$\text{Cost per m}^2 = \frac{\text{Building cost}}{\text{Total floor area}}$$

$$= \frac{£16,600,000}{1500} = £11,000.00 \text{ approximately}$$

---

### Cube method

In the cube method the cubic content of a building, that is the volume displaced by air, is calculated in accordance with the RIBA rules of measurement. This cubic content is multiplied by a rate per $\text{m}^3$ in order to establish the estimated cost.

The cube method has now been replaced by other more accurate methods of approximate cost estimates. It is felt that the volume of air displaced within the building does not directly reflect the quantities of both labour and materials forming the enclosure, the vertical and the horizontal divisions of a structure.

## Storey enclosure method

The storey enclosure of single price-rate estimating aims to take the following factors into account:

- Shape of the building
- Total floor area
- Vertical positioning of the floor areas
- Storey heights of buildings.

This method has been rarely used in practice, mainly because it involves more calculations than the methods described previously, and also because there are no rates published for this method.

## Elemental method

The elemental method provides a method of examining and comparing building costs using information from a Bill of Quantities. The contract sum or tender figure is subdivided into the costs of the various building elements, as set out in the rules prescribed by the RICS, BCIS standard form of cost analysis.

These elemental costs are then expressed by means of a common criterion, such as the cost per m² of gross floor area.

*An **element**: that part of a building which always performs the same function, irrespective of the type of building.*

### Advantages of elemental method

- Rapid and accurate method of preparing realistic estimates
- Allows for adjustments to be made to individual elements
- Excellent base system for making comparisons with the cost distribution for the individual elements
- Also indicates possible cost distribution for the individual elements in addition to lump sum figure.

### Disadvantages of elemental method

- A degree of skill and understanding of cost planning is required for interpreting the difference between the elements, and calculating their effect on cost
- Method requires more time to prepare than others.

## Approximate quantities

The method of approximate quantities is generally considered to be the most reliable and accurate method of approximate estimating. It involves more work than the other methods and, on occasions, lack of information precludes its use. Maximum use is made of one set of dimensions for a number of items or work sections.

For example, an upper floor price rate may consist of all the following work elements 'lumped together' into one rate per $m^2$ of floor:

- Floor construction
- Ceiling finish and decoration
- Floor screed and finish
- A percentage addition to cover smaller items such as labour, skirting and decorations.

### Advantages of approximate quantities

- An estimate can be divided into individual items or elements which assist the development of a balanced design
- Adjustments can be made to an estimate for specification changes and other reasons with the minimum of fuss or disturbance.

---

**Example of approximate quantities estimate**

*Store building*

| | | | |
|---|---|---|---|
| Excavate oversite; lay hardcore; polythene membrane and 150 mm concrete | 20 m² | 40.00 | 800.00 |
| 300 mm thick cavity wall construction; faced externally with facing bricks; two coats of paint internally | 40 m² | 90.00 | 3,600.00 |
| Extra over for metal windows, including glazing and ironmongery | 6 no. | 75.00 | 450.00 |
| Extra over for doors as above | 2 no. | 60.00 | 120.00 |
| Reinforced concrete roof, average 150 mm thick; reinforced and covered with two coats of asphalt | 18 m² | 90.00 | 1,620.00 |

*Add for*:

| | |
|---|---|
| Electrical works | 610.00 |
| Telephone | 300.00 |
| Preliminaries | 1,000.00 |
| Contingencies | 1,000.00 |
| Estimated cost of building | 9,500.00 |

## Elemental cost planning

This process of elemental cost planning represents the 'cost estimate', which is subdivided into various elements of construction, with each element allocated a cost based on previously constructed similar projects. This technique is often referred to as *designing to a cost*, because it originates from scheme design.

## Comparative estimating

The method of comparative estimating also originates from scheme design, but does not use a fixed budget like the elemental system. Here a cost study is made which shows the various ways in which the design may be performed and gives the cost of each alternative approach. The cost study indicates where the project can be carried out within the cost limit laid down by the client. This process is often referred as *costing to a design*.

## Interpolation method

The interpolation method is a variant of the comparative method, whereby an estimate of probable cost is produced by taking the cost per square metre of floor area of a number of similar type buildings from cost analysis and cost records, and interpolating a unit rate for the proposed building.

The simplistic nature of the method can cause problems, as no two buildings are the same. It is difficult to make adjustments to the unit rate to take account of the many variables that are bound to occur between the buildings. In practice, it may often be necessary to use a method which is a combination of both the interpolation and comparative approaches.

## Elemental cost analysis

Elemental cost analysis uses the systematic breakdown of cost data, often on the basis of elements, to assist in estimating the cost and in the cost planning of future projects. Cost analyses are often supplemented with specification notes, data concerning site and market conditions and various quantity factors, such as wall-to-floor ratios. The format may be in a concise or a detailed format.

**Typical contents of a cost analysis report**
Job title
Location
Client details
Tender date
Project details
Site conditions
Contract type
Market conditions
Basis of tender
Cost fluctuations
Provisional sums
Prime cost sums
Contingencies
Contract sum
Competitive tender lists
Tender sums
Elemental cost breakdown
   (6 elements)

## Cost modelling

***Cost modelling** is the symbolic representation of a system in terms of the factors which influence its cost.*

Cost models attempt to represent the significant cost items of a cashflow, or building or component, in a form which will allow analysis and prediction of costs. The model will then show the effects of changes on design variables, methods of construction, timing of events and other factors.

The principle of any modelling technique is to simulate a product by building a prototype. Cost modelling adopts a number of alternative mathematical formats or simulations to generate results which may then be analysed and used in the design-making process. The technique is used widely in preparing cost estimates for construction work.

Examples of the more traditional cost modelling techniques were discussed earlier in the chapter and a summary of modelling techniques is listed in the margin.

---

**Keywords for cost planning**

The following is a list of some keywords used in this chapter. Use the list to test your knowledge and, if necessary, consult the text to learn about the terms.

| | | |
|---|---|---|
| Approximate estimating | Comparative estimating | Cost plan |
| Elemental cost planning | Interpolation | Cost checking |
| Unit method | Elemental cost analysis | |
| Approximate quantities | Cost modelling | |

# 8 Tendering Methods and Cost Control

This chapter is concerned with the later design stages of the RIBA Plan of Work which is described in Chapter 6. The main methods of tendering are described and lead to an appreciation of the financial processes used when the contract commences and works begin on site.

## Tendering methods

The object of tendering is to select a suitable contractor, at the proper time, and to obtain an acceptable offer to execute construction works. Prior to selecting a suitable tendering method, the following factors should be taken into account:

- Size of project
- Character of project
- Location of project
- Level of pricing
- Period of time for construction project
- Type of firm and type of business carried out
- Financial resources of firm
- Physical resources of firm
- Human resources of firm
- Past performance of firm
- National economic climate.

The following are the principal tendering methods available:

- Open tendering
- Restrictive open tendering
- Selective tendering
- Negotiated tendering
- Serial tendering
- Two-stage tendering.

## Open tendering

The press often carries advertisements which are open invitations for any contractor to apply for a set of tender documents. A small deposit is required, returnable on receipt of a genuine tender.

---

*Advantages of open tendering*

- Very competitive tenders are obtained
- Only interested firms will submit tenders
- New firms are able to obtain work and prove themselves.

*Disadvantages of open tendering*

- Some firms may not be well-equipped, either materially or financially, to execute the work
- If a very low tender is submitted and accepted, it may cause difficulties throughout the contract
- Submitting tenders costs time and expense, and this cost needs to be recovered.

---

## Restrictive open tendering

The method of restrictive open tendering also uses advertisements which invite applications from firms that wish to be considered for tendering. From these applicants a selection list is chosen and tenders are invited.

---

*Advantages of restrictive open tendering*

- Only interested firms will apply
- Only suitable firms are asked to submit tenders
- Less overall expense for tenderer
- New firms may be able to obtain work and prove themselves.

*Disadvantages of restrictive open tendering*

- Less competitive
- Can lead to cover pricing being submitted.

---

## Selective tendering

In selective tendering the Architect prepares a list of suitable tenderers, and they alone are invited to submit a tender. The tender list normally contains 6 to 8 firms, according to the size and type of the proposed project.

---

*Advantages of selective tendering*

- Only firms capable of executing the work will be selected
- Selected firms will probably have already proved themselves
- Reduction in the time and overall cost of tendering.

*Disadvantages of selective tendering*

- The price may not be as competitive
- Can lead to cover prices being submitted
- Difficult for new firms to obtain work easily.

---

## Negotiated

Tendering by negotiation involves the selection of a single firm for the purpose of discussing the project and eventually agreeing a price. The firm selected may be based on the general criteria outlined earlier, possibly after an initial interview with the Architect when the method of execution and other factors are discussed.

Tender documents are prepared in the normal manner and sent to the selected contractor who prices the work. When the final tender figure has been prepared it is submitted to the Quantity Surveyor for examination and a tender report. The contractor and the Quantity Surveyor will meet to settle any queries and adjust details such as unreasonable rates. After negotiation, a price that is acceptable to both parties is agreed and a formal contract is signed.

---

*Advantages of negotiated tendering*

- Contractor can be selected early in the design stage and can assist in the contract planning
- Useful for a job of a difficult or unusual nature
- Useful for extension contracts

*(continued)*

---

*Advantages of negotiated tendering* (*continued*)

- Useful where there is insufficient time to prepare full tender documentation
- Establishes good relationships between client and contractor
- Low overall cost of tendering.

*Disadvantages of negotiated tendering*

- The price may be higher than in open competition
- The design could be influenced by the selected contractor
- Difficult for new firms to obtain work.

---

## Serial tendering

Serial tendering is a form of standing offer where a contractor undertakes to enter into a series of separate fixed price contracts in accordance with terms and conditions set out in the standing offer. The offer usually relates to a firm programme of projects of a similar nature within reasonable graphical limits, such as a programme of building supermarkets.

Preliminary discussions are held with a short list of contractors in order to draw up conditions of the offer, such as the number of projects, the time period and the phasing of jobs. A notional Bill of Quantities is used to help evaluate the offers.

When the offer is accepted, Bills of Quantities are prepared for each project which are then priced in accordance with the notational Bill of Quantities and an agreed tender negotiation. The initial offer is merely a statement made in good faith as a basis for negotiation that may be withdrawn at any time.

---

*Advantages of serial tendering*

- A competitive price may be obtained
- The contractor is able to plan a long-term programme
- The experience obtained in earlier projects may be used to the advantage of both parties for the future projects.

*Disadvantage of serial tendering*

- If the contractor initially produces unsatisfactory work the client may be committed to a long-term programme.

## Two-stage tendering

There are occasions, such as those listed below, when selection of the most appropriate contractor is of prime importance:

- The client wishes to start work on site early and prior to full tender documentation being prepared
- The contractor can make a technical contribution to the project.

If the drawings and other details are not yet finalised then the client may invite contractors to tender for a contract after an initial and careful negotiation with each contractor about how the proposed project should be undertaken. The selected contractors are then invited to competitively tender for the work by pricing one of the following items:

- Approximate Bill of Quantities
- Bill of Quantities of a similar past project
- A fictitious Bill of Quantities.

The contractor who submits the most favourable price (tender) is then expected to work closely with the designer to agree on an economic design and programme until a satisfactory solution to the client's needs is realised. A final tender is then submitted by the selected contractor using the Bill prices previously outlined in the successful competitive tender.

---

*Advantages of two-stage tendering*

- Speed in planning and construction
- Use of contractor's specialist design team and solutions
- Use of the contractor's management and site-solving solutions.

*Disadvantages of two-stage tendering*

- High overheads of unsuccessful first-stage contractor
- High costs of variations.

---

# Post-contract processes

The financial processes that occur after the contract has commenced are known generally as post-contract financial processes. The following are important post-contract processes:

- Interim valuations
- Financial statements
- Final accounts
- Claims.

## Financial reports

**Financial report items**
Contract sum
Contingencies
Variations
Fluctuations
Claims

As the project progresses, the Quantity Surveyor should keep the building owner (client) informed about the financial position of the contract. A useful report includes information on the known and anticipated expenditure compared with the overall budget allowances. This report often accompanies the Quantity Surveyor's interim valuation recommendation.

The building owner is concerned not only with the current cost position, but also with the likely pattern of future payments over the remaining contract period. This information would normally be presented in the form of an updated cashflow forecast by which the building owner can ensure that there are sufficient funds to pay for the balance of the projected final costs.

In the event of substantial additional expenditure, it may be necessary for the building owner to raise further finance. An alternative may involve the building owner requesting that savings are made in the project design in order to keep the final costs within the budget figure.

Many of the disputes which arise on construction contracts could be avoided by better communication and the adoption of formalised procedures for the regular reporting on the financial state of the works to the parties.

---

### Some contract terminology

*Contract sum*

Price for carrying out construction as entered on the contract documents.

*Contingencies*

A sum inserted into the contract documents to cover alterations to the project without the need to approach the client for additional funds.

*(continued)*

## Some contract terminology (*continued*)

### Fluctuations

A provision in a building contract to compensate a contractor for increases in the cost of labour and material resources during the course of the project.

### Claims

An entitlement to additional payment to the contractor for delay and disruption to the contract works.

## Interim valuations

It is standard practice in construction contracts for contractors to receive payment from the building owner at regular intervals, usually monthly, as the project progresses. This payment is commonly referred to as an interim certificate or interim valuation.

Normally, it is the responsibility of the building owner's consultant Quantity Surveyor to prepare the valuation and recommend an amount due to the contractor. The Architect will normally accept the Quantity Surveyor's valuation statement and issue an *interim certificate* which is, in turn, forwarded to the building owner as authority to make payment to the contractor.

The degree of accuracy required when preparing an interim valuation must be balanced by the fact that the Quantity Surveyor must safeguard the interests of the building owner, yet also be fair and legal towards the contractor.

The components of an interim valuation may include:

- Value of works executed:
    Preliminaries
    Measured work
    Nominated subcontractors' work
    Nominated suppliers' work
    Profit and attendance for the above
- Value of variations
- Value of dayworks
- Value of unfixed materials on site and off site
- Value of any contractual claims
- Value of fluctuations
- Value of retention held and released
- Previous payments to contractor
- Recommended value of certificate.

The degree of measurement to be undertaken depends upon the nature and complexity of the works and the stage they have reached. It is common practice for the Quantity Surveyor and the contractor to meet at regular intervals, usually monthly, for the purpose of valuing the works.

---

### Example of an interim valuation

| | |
|---|---:|
| Preliminaries | 28,000.00 |
| Measurement work | 185,000.00 |
| External works | 20,000.00 |
| Nominated subcontractors/suppliers | 13,000.00 |
| Variations | 1,800.00 |
| Materials on site | 13,000.00 |
| Materials off site | 7,000.00 |
| Claims | 10,000.00 |
| Fluctuations | 5,000.00 |
| | 282,800.00 |
| *Less* | |
| Retention @ 3% | 8,034.00 |
| (not in claims or | |
| fluctuations) | 274,766.00 |
| *Less* | |
| Previous payment | 196,000.00 |
| **Interim payment due** | 78,766.00 |

---

### Final accounts

The final account is the final statement or claim prepared by the consultant Quantity Surveyor in conjunction with the contractor. The account should include all claims in respect of all work carried out by the main contractor and any specialist subcontractors or suppliers.

The contractor should forward to the Quantity Surveyor all the necessary information in order for the Quantity Surveyor to prepare the account. When the final account has been completed a copy is sent to the contractor, together with extracts to any nominated suppliers or subcontractors, for their agreement.

The bulk of the final account will consist of measured work priced at rates of the original Bill of Quantities. However, a number of 'adjustments' must be made to the original contract sum, as listed.

When preparing the final account, the Quantity Surveyor should allow the contractor the opportunity to be present when measurements or details are recorded. The draft final account is a useful mechanism in maintaining cost control of the contract if it is commenced at the beginning of the project and is updated as the project progresses.

**Adjustments for final account**
Variations
Remeasurement
Nominated
   subcontractors
Nominated suppliers
Claims
Fluctuations

---

**Example of a final account**

| | |
|---|---:|
| Total amount of original contract sum | 278,000.00 |
| *Less* | |
| Provisional sum adjustment | 35,000.00 |
| | 243,000.00 |
| *Less* | |
| Prime sum cost adjustment | 40,000.00 |
| | 203,000.00 |
| *Add* | |
| Expenditure on provisional and prime cost sums | 70,000.00 |
| | 273,000.00 |
| *Add* | |
| Variations | 50,000.00 |
| *Add* | |
| Fluctuations adjustment | 10,000.00 |
| *Add* | |
| Recovery on claims | 15,000.00 |
| | 348,000.00 |
| *Add* | |
| Value added tax @ 17.5% | 60,900.00 |
| **Final account sum** | 408,900.00 |

---

## Claims

*A **claim** is a request by a contractor for additional payment to which he believes he is entitled by the terms of the construction contract.*

In general, the contractor is required to give notice on the occurrence of certain events. Contractual claims arise from unexpected events which can be grouped as follows:

**Claims settlement**
*Investigation*: to prove that delay has occurred and that damage has been suffered.
*Assessment*: to evaluate the value of the loss or expense.

- Issue of variations in the specifications and quantity of works
- Disturbance of the regular progress of the works.

Settlement of claims is not a precise science but uses the basic principles of investigation and assessment. The preparation of a claim involves consideration and possible use of the following headings:

- Materials
- Labour disruption
- Attraction money and bonus payment
- Preliminaries and supervision
- Inflation
- Head office overheads and profit
- Interest charges
- Cost of accelerating the works
- Cost of overtime
- Cost of preparing a claim
- Out-of-sequence working.

---

**Keywords for cost control**

The following is a list of some keywords used in this chapter. Use the list to test your knowledge and, if necessary, consult the text to learn about the terms.

| | | |
|---|---|---|
| Open tendering | Two-stage tendering | Investigation |
| Selective tendering | Interim certificate | Assessment |
| Serial tendering | Final account | |
| Open restrictive tendering | Claim | |

# 9 Measurement Standards

The extent of the work involved in a construction project will vary from project to project. Some projects will involve a small number of basic activities, while other larger projects may require a large number of more complex activities. We have seen in earlier chapters how these differing projects can be resourced and programmed.

In all building and civil engineering projects there is a need for a common approach and language in the measurement of construction work. This measurement is necessary in order to determine accurately the probable cost of a future construction project. The skill of measurement is normally carried out by a Quantity Surveyor.

Measurement involves the Quantity Surveyor in measuring different types of work as shown on the drawings produced by the Architect or engineer. The quantities are prepared in accordance with the rules of a domestic or international measurement code and the tender documentation prepared is commonly referred to as a *Bill of Quantities*. When completed, the Bill of Quantities is normally forwarded, along with other documentation, for the contractor to price.

The task of measurement is also used while a building or engineering project is under construction. This process enables the Quantity Surveyor to measure work as the work proceeds and to value the work on a monthly basis for payment purposes. Measurement is also used for the preparation of any claims, as well as the settlement of the final account.

Table 9.1 identifies the role of measurement in the overall construction process.

## Standard approach

Over the years, the industry has developed measurement codes which have become recognised as acceptable rules in the measurement of construction works.

**Table 9.1**   Role of measurement*

| Outline construction process | Role of measurement |
| --- | --- |
| Briefing | Not applicable |
| Sketch plan | Measurement of approximate quantities for the purpose of preparing approximate estimates and cost checking exercises |
| Working drawings | Taking off quantities from drawings and preparation of Bill of Quantities |
| Site operations | Measurement of work on site for the purpose of preparing valuations, payments to subcontractors and suppliers, claims, and final account |

* Adapted from RIBA plan of work for design team operations.

The principal reasons for having consistent rules of measurement are as follows:

- Provides an industry standard
- Provides a common language for measurement
- Represents all sectors of the industry
- Allows for the production of standard tender documents
- Allows pricing on an equal basis
- Allows for uniformity in costing
- Reduces risk to contractor and client.

# Standard methods of measurement

The first measurement code in the United Kingdom was published in 1922 and provided at the time an outline guide for the measurement of buildings.

To date, the industry has progressed to the extent that we now have measurement codes for building, civil engineering, industrial and other categories of construction activity.

This chapter will cover the two most widely used and accepted codes of measurement in the United Kingdom which are:

- The Standard Method of Measurement for Building Works 7th Edition (SMM7).
- The Civil Engineering Standard Method of Measurement 3rd Edition (CESMM3).

The standard method of measurement for building works is issued by the RICS (The Royal Institution of Chartered Surveyors) and the BEC (Building Employers Confederation) and forms the basis for the measurement of the bulk of building works in the United Kingdom, and in other countries.

The SMM7 document was published in January 1988, together with the accompanying *Code of Measurement Practice Manual* which was the successor to the SMM6 practice manual. Both documents came into use in July 1988.

The Practice Manual explains and enlarges upon the main measurement code as necessary and gives guidance on the interpretation of certain measurement rules in each work section. It is important to note that the Practice Manual is for guidance only and does not have the mandatory status of the SMM7 document.

The standard method of measurement for civil engineering works is issued by the ICE (Institute of Civil Engineers) and the FCEC (Federation of Civil Engineering Contractors) and forms the basis for the measurement of the majority of civil engineering projects. The first edition was published in 1974 and today the third edition is currently in use.

The publication of measurement codes is widely adopted internationally. An international measurement code known as POMI/1 (Principles of Measurement – International Edition 1) is available for larger international infrastructure projects such as harbours, roads and airports.

**Table 9.2** Countries and standard methods

| Country | Standard method of measurement for building work |
| --- | --- |
| Belgium | Code de mesurage |
| Finland | Building 80 |
| France | 1. Mode de métré normalisé – Avant<br>2. Métré et devis quantitatif des ouvrages de bâtiment |
| Germany | VOB/C – Verdingungsordnung für Bauleistungen – Part C |
| Great Britain | Standard Method of Measurement (7th edition) |
| Ireland | Agreed Rules of Measurement<br>Standard Method of Measurement 6th Edition |
| Netherlands | 1. Nen 3699<br>2. Standard Measurement of Net Quantities for Materials and Activities of Building |

The use of measurement codes is also evident in the countries of the European Union. Table 9.2 lists the principal codes used in a selection of countries.

# The SMM7 Measurement Code

The Standard Method of Measurement Code explains the rules which must be followed when measuring different types of work in construction.

**Table 9.3** General rules

| General rules | Coverage in brief |
| --- | --- |
| Introduction | Emphasis is on the uniform basis of the rules and the need for additional information if required |
| Use of tabulated rules | Describes the structure and use of the classification tables and supplementary rules |
| Quantities | Highlights the main rules in connection with the entries of quantities to dimension sheets |
| Descriptions | Confirms order of dimensions, information required and items deemed to be included |
| Drawn information | Identifies the typical drawn information that should be available for the take-off |
| Catalogued or standard components | Use of unique cross-reference to a catalogue or a standard specification in lieu of a detailed description |
| Work of special types | Identifies special types of work which must be separately identified when taking-off |
| Fixing, base and background | Identifies the need for clarity in describing the nature of the fixing, base and background for Bill of Quantity items |
| Composite items | Considers work items normally manufactured off-site and identifies items deemed to be included |
| Provisional sums | Considers work not entirely identified or quantified at the time of preparing the Bill of Quantities |
| Work not covered | Covers items not considered in the Rules and calls for a consistent and compatible approach |
| Work to existing | Work to existing buildings is described |
| General definitions | Clarifies the position in respect of curved work |

The SMM7 rules are structured in the following parts:

- General Rules
- Common Arrangement of Work Sections
- Additional Rules
- Appendices.

## General Rules

These rules apply on a uniform basis for measuring all work sections under the measurement code and embody the essentials of good measurement practice. Table 9.3 gives a brief account of the rules covered in this section.

General rule 3.0 identifies the following rules in connection with calculating quantities:

- Work shall be measured net as fixed in position
- Dimensions shall be given to the nearest 10 mm
- Quantities measured in tonnes shall be given to two decimal places
- Any quantities less than one unit shall be given as a whole unit
- Minimum deduction for voids shall refer to voids only within the boundary of the area.

A general rule identifies key symbols and abbreviations, which are shown in Table 9.4.

## Common Arrangement of Work Sections

The bulk of the Standard Method concerns the rules of measurement for individual work sections in a common arrangement order. The structure of the SMM7 differs in many ways to its predecessor SMM6. The first major change was from trade format to the order of the Common Arrangement of Work Sections.

The *Common Arrangement of Work Sections* (CAWS), shown in Table 9.5, is based on the natural groupings as devised by the Coordinating Committee for Project Information (CCPI) which was established in 1979 by RIBA (Royal Institute of British Architects), the RICS (Royal Institution of Surveyors) and the BEC (Building Employers Confederation). The objective of the arrangement was to bring about a general improvement in the documents and communications used for the design, procurement and construction of buildings.

**Table 9.4**

| Symbol | Meaning |
| --- | --- |
| m | metre |
| $m^2$ | square metre |
| $m^3$ | cubic metre |
| mm | millimetre |
| nr | number |
| kg | kilogram |
| t | tonne |
| h | hour |
| pc | prime cost sum |
| > | exceeding |
| $\geq$ | equal to or exceeding |
| $\leq$ | not exceeding |
| < | less than |
| % | percentage |
| – | hyphen |

**Table 9.5**   Common Arrangement of Work Sections (CAWS)

| Reference | Work section |
|-----------|--------------|
| A | Preliminaries/General conditions |
| B | General rules |
| C | Demolition/Alteration/Renovation |
| D | Groundwork |
| E | *In-situ* concrete/Large precast concrete |
| F | Masonry (Brickwork, blockwork and stonework) |
| G | Structural/Carcassing Metal/Timber |
| H | Cladding/Covering |
| J | Waterproofing |
| K | Linings/Sheathing/Dry partitioning |
| L | Windows/Doors/Stairs |
| M | Surface finishes (plastering, tiling, painting etc.) |
| N | Furniture/Equipment |
| P | Building fabric sundries |
| Q | Paving/Planting/Fencing/Site furniture |
| R | Disposal systems |
| S | Piped supply systems |
| T | Mechanical heating/Cooling/Refrigeration systems |
| U | Ventilation/Air conditioning systems |
| V | Electrical supply/Power/Lighting systems |
| W | Communications/Security/Control systems |
| X | Transport systems (lifts, escalators etc.) |
| Z | Mechanical and electrical services |

**Preliminaries/General conditions**

Project particulars
Drawings
The site/existing buildings
Description of the work
The contract/subcontract
Employer's requirement
Contractor's general cost items
Work/materials by the employers
Nominated subcontractors
Nominated suppliers
Work by statutory authorities
Provisional work
Dayworks

Another major feature of the SMM7 is the adoption of a tabulated format as opposed to the prose of the previous edition. This change is largely adapted from the successful *Civil Engineering Code of Measurement 2nd Edition*. The document is also presented on the page in landscape (sideways) rather than portrait format, in order to achieve a more acceptable and useful layout.

There are twenty three common arrangements (see Table 9.5) which can be conveniently divided into two areas:

- Preliminaries/General conditions
- Work sections.

The section on preliminaries and general conditions covers the general introductory information relevant to all projects and this is set out in common arrangement order. Each topic has a common arrangement reference such as A10 Project Particulars, A20 The Contract, and A54 Provisional Work. A list of the areas covered in this section is shown in the margin.

The work sections of SMM7 concern the rules of measurement for the various items of work typically encountered in the construction of buildings. Each work section, including the section on preliminaries/general conditions, contains rules which may be divided under the following three headings:

- Information required
- Classification table
- Supplementary rules.

Table 9.6 gives a brief account of the basis of the rules.

The additional rules in the SMM7 document are for use in connection with work to existing buildings. The Appendix section of the document concentrates on the scope of particular work sections.

The rules prescribed in the Standard Method of Measurement Code are set out under each work section and organised in a tabular form made up of *classification tables* and *supplementary rules*. Horizontal lines divide the classification table and rules into distinct zones in which different rules apply (see Figure 9.1).

**Table 9.6** Work Section Rules for SMM7 Code

| *Work section headings* | *Coverage* |
| --- | --- |
| Information required | Information which should be provided in the measured work section as an aid to tendering |
| Classification table | Lists descriptive features commonly encountered in building works. They are further subdivided by a tabular format which allows preparation of a structured description for entry on to the dimension paper |
| Supplementary rules | |
| a) Measurement rules | Set out when work will be measured and the method by which quantities shall be computed |
| b) Definition rules | Define the extent and limits of the work represented by a word or expression used in the rules and in a Bill of Quantities |
| c) Coverage rules | Draw attention to particular incidental work deemed to be included in the appropriate items in a Bill of Quantities |
| d) Supplementary information | Contains rules governing the extra information which shall be given |

| CLASSIFICATION TABLE | | | | MEASUREMENT RULES | DEFINITION RULES | COVERAGE RULES | SUPPLEMENTARY INFORMATION |
|---|---|---|---|---|---|---|---|
| 1 Site preparation | 1 Removing trees<br><br>2 Removing tree stumps | 1 Girth 600 mm to 1.50 m<br><br>2 Girth 1.50 to 3.00 m<br><br>3 Girth > 3.00 m, girth stated | nr | M1 Tree girths are measured at a height of 1.00 m above ground<br>M2 Stump girths are measured at the top | | C1 This work is deemed to include:<br>(a) grubbing up roots<br>(b) disposal of materials<br>(c) filling voids | S1 Filling material described |
| | 3 Clearing site vegetation | 4 Description sufficient for identification stated | m² | | D1 Site vegetation is bushes, scrub, undergrowth, hedges and trees and tree stumps ≤600 mm girth | | |
| | 4 Lifting turf for preservation | 1 Method of preserving, details stated | m² | | | | |

**Figure 9.1**   Typical layout of Standard Method of Measurement document.

Where broken horizontal lines appear within a classification table, the rules entered above and below the lines may be used as alternatives. Within the supplementary rules, all information above the first horizontal line, which is immediately below the classification table heading, is applicable throughout the table.

The left-hand column of the work section classification table lists descriptive features commonly encountered in building works, followed by the relevant unit of measurement. The second column lists subgroups into which each main group can be divided and the third column provides a further subgrouping. Each item description identifies the works by drawing a feature from each of the first three columns in the classification, and as many of the features in the fourth or last column as are appropriate.

The SMM7 code allows the surveyor to cross-reference each description to the classification table columns in each work section (see Figure 9.1 and Table 9.7). Additional rules in relation to cross-referencing include the following:

- An asterisk within a cross-reference represents all entries in the column in which it appears
- The digit 0 within a cross-reference represents no entries in the column in which it appears.

**Table 9.7** Example of cross-referencing in SMM7 Code

| Classification levels Work section | SMM7 reference | Description levels |
|---|---|---|
| Work section code | D20 | Excavation and filling |
| Numbered from first column | 1 | Excavating |
| Numbered from second column | 2 | To reduced levels |
| Numbered from third column | 2 | Maximum depth $\leq 1.00\,\mathrm{m}$ |

# Presentation of quantities

The presentation of quantities is standardised for the purposes of good practice. The measured quantities are presented on standard ruled paper. The dimensions are either scaled from the drawings, or calculated directly from the drawing and entered into the appropriate column on the dimension sheet.

### Dimension paper

The standard dimension sheet is A4 in size (210 mm × 297 mm) and printed in portrait format (see Figure 9.2 on page 130). The format conforms with the requirements of BS 3327 1970: Stationery for Quantity Surveying.

| Key to dimension paper | | |
|---|---|---|
| Column | Title | Function |
| 1 | Binding | Column kept clear – to allow papers to be bound together |
| 2 | Timesing | Multiplying figures – if there are more than one of the particular items being measured |
| 3 | Dimension | The actual dimensions – as scaled or taken direct from the drawings. There may be one, two or three lines of dimensions in an item depending on whether it is a linear, square or cubic measurement respectively. All dimensions are entered to two decimal places |

(*continued*)

**Quantity surveying stationery**
Single bill headed
Single bill
Double bill
Estimating single bill
Right hand single bill
Right hand double bill
Abstract paper

**Key to dimension paper** (*continued*)

| Column | Title | Function |
|---|---|---|
| 4 | Squaring | Length, area or volume – obtained by multiplying together the figures in columns 2 and 3. This figure is ready for transfer to the abstract or the bill later in the measurement process |
| 5 | Description | Written descriptions of each item. Waste calculations can be shown in the right-hand portion of column. Annotations of this can be shown on the left-hand portion of this column |

### Cut and shuffle

An alternative format is to use a *cut and shuffle* method. This format is used when the Bill of Quantities is prepared using the cut and shuffle method of taking-off and bill production which is described in more detail in Chapter 11.

Cut and shuffle paper is available in 2, 3, 4 or 6 slip format.

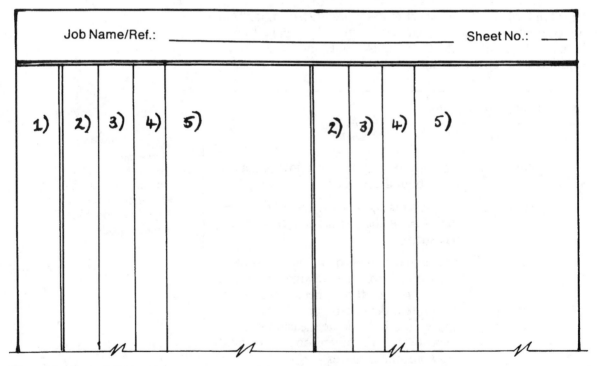

**Figure 9.2** Typical dimension paper.

# Measurement practice

The process of measurement is commonly referred to as 'taking-off'. When preparing the take-off the surveyor should possess certain skills. A checklist is shown in the margin.

Good measurement practice involves the following stages:

1. Read and check the drawings and specifications
2. Prepare calculations sheet
3. Prepare take-off list/plan
4. Prepare query sheets
5. Enter dimensions
6. Commence take-off
7. Check and double check all work.

It is important that the following qualities are present in measurement take-off:

- Neatness of presentation
- Good taking-off principles
- Accuracy of dimensions.

**Taking-off skills**
Sound mathematical knowledge
Ability to read construction drawings
Ability to envisage details not shown
Ability to interpret tender documents
Attention to detail
Neatness of presentation
Accuracy and relative speed
Good measurement practice

**Table 9.8** Typical content of drawings

| Drawings | Type | Content |
|---|---|---|
| Location drawings | Block plan | Identifies the site and locates the outline of the building works in relation to a town plan or other context |
| | Site plan | Locates the position of the building works in relation to the setting out points, means of access, and general layout of the site |
| Plans, sections and elevations | | Show the position occupied by the various spaces in a building and the general construction and location of the principal elements |
| Component drawings | | Show the information necessary for manufacture and assembly of a component |
| Dimensioned diagram | | Shows the shape and dimensions of the work covered by an item and may be used in a Bill of Quantities in place of a dimensioned description |

## Drawn information

The drawn information used by the Quantity Surveyor may take the following forms:

- Location drawings
- Component drawings
- Bill diagrams.

## Query sheets

**Typical queries**

Any trial holes or ground condition information?

Any existing services on or above the site?

Confirm existing site level

Confirm specification of facing bricks

Confirm pre-contract ground water level

Preliminary inspection of the drawings may highlight a number of questions which need answers from the Architect or engineer. It is good practice to enter these queries on a prepared schedule so that the Architect or engineer can deal with all the queries together (see Figure 9.3).

The finding of answers to such queries at an early stage saves interruption of the taking-off process and increases the productivity in achieving the end product.

## Take-off list

The order of taking-off generally follows the order of construction. The items are usually arranged in a bill order which follows the work sections in the standard method of measurement.

---

*Typical take-off list for substructure*

Excavate topsoil and disposal
Excavate trench and disposal
Extra over excavation for ground water
Disposal of surface and ground water
Earthwork support
Foundation concrete
Formwork (if any) to foundation concrete
Hardcore and surface treatment
Brickwork
Damp-proof course
Concrete bed
Damp-proof membrane
Adjustment of excavation backfilling
Adjustment of topsoil backfilling

---

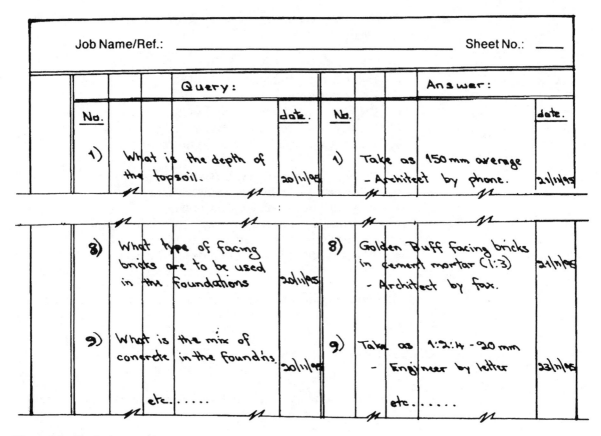

**Figure 9.3**   Typical query sheet.


## Entering dimensions

Dimensions are entered on to the dimension paper in one of the
following five forms:

- Number or enumerated
- Linear measurement
- Square or superficial measurement
- Cubic measurement
- Item.

For the first three forms it is good practice to set down
dimensions in the following order:

- Horizontal length
- Horizontal width or breath
- Vertical depth or height.

When entering dimensions, the following guidelines should be followed:

- Enter project reference and title
- Enter page number
- Provide clear side notes and annotations
- Clearly label side calculations
- Restrict waste calculations to description column only
- Provide thumbnail sketches when appropriate
- Allow adequate spacing between descriptions
- Use brackets where appropriate
- Order of dimensions is: length, width and depth
- Dimensions to two decimal places
- Waste calculations to three decimal places.

The above guidelines are illustrated in Figures 9.4 and 9.5.

### Taking-off summary

Use the following checklist to help achieve satisfactory results when taking-off quantities:

- Study drawing to obtain a clear picture of what you are to measure
- Check dimensions on drawing
- Decide upon section or part of the works you are going to measure first
- Prepare query sheets
- Prepare take-off list/plan
- Study the rules of the relevant work sections
- List item headings as memory aid
- Commence take-off
- Ensure good measurement practice at all times
- Mark/tick items on the take-off list/plan when measured
- Tick all notes on drawing as you measure
- Scan drawings for any unmeasured items
- Check all side calculations.

**Reminder**: Aim to achieve a balance between the skills of presentation, approach and accuracy while carrying out the measurement processes.

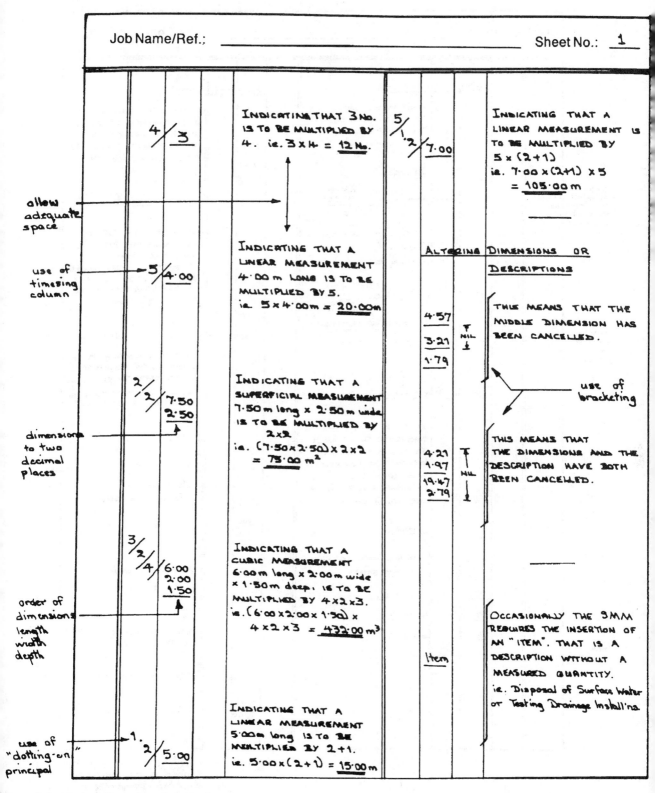

**Figure 9.4** Typical dimension sheet entries.

135

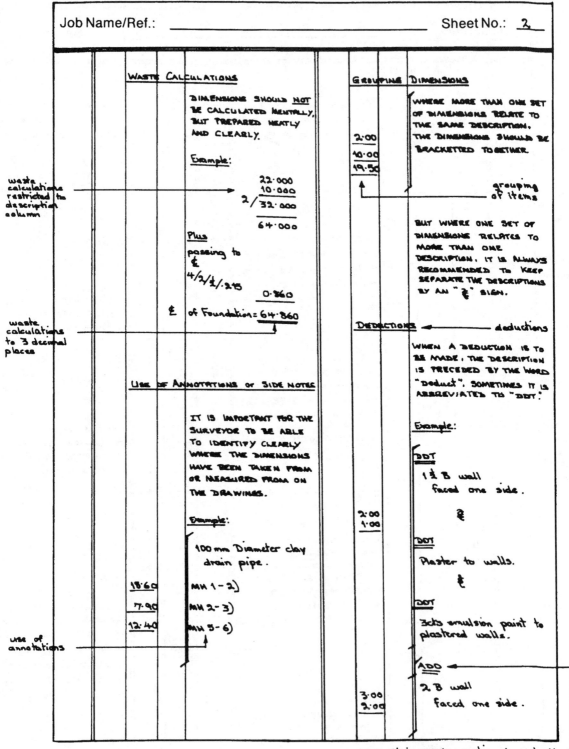

WASTE CALCULATIONS

DIMENSIONS SHOULD **NOT** BE CALCULATED MENTALLY, BUT PREPARED NEATLY AND CLEARLY.

Example:

```
            22·000
            10·000
        2 / 32·000
            _____
            64·000
```

Plus
passing to
£
4/2/½/·215        0·860
                  _____
£ of Foundation = 64·860

*waste calculations restricted to description column*

*waste calculations to 3 decimal places*

USE OF ANNOTATIONS OR SIDE NOTES

IT IS IMPORTANT FOR THE SURVEYOR TO BE ABLE TO IDENTIFY CLEARLY WHERE THE DIMENSIONS HAVE BEEN TAKEN FROM OR MEASURED FROM ON THE DRAWINGS.

Example:

100 mm Diameter clay drain pipe.

18·60   MH 1 - 2)
7·90    MH 2 - 3)
12·40   MH 5 - 6)

*use of annotations*

GROUPING DIMENSIONS

WHERE MORE THAN ONE SET OF DIMENSIONS RELATE TO THE SAME DESCRIPTION, THE DIMENSIONS SHOULD BE BRACKETTED TOGETHER

2·00
10·00
19·50

*grouping of items*

BUT WHERE ONE SET OF DIMENSIONS RELATES TO MORE THAN ONE DESCRIPTION, IT IS ALWAYS RECOMMENDED TO KEEP SEPARATE THE DESCRIPTIONS BY AN "&" SIGN.

DEDUCTIONS ← *deductions*

WHEN A DEDUCTION IS TO BE MADE, THE DESCRIPTION IS PRECEDED BY THE WORD "Deduct". SOMETIMES IT IS ABBREVIATED TO "DDT".

Example:

DDT
1½ B wall
    faced one side.
&

2·00
1·00

DDT
Plaster to walls.
&

DDT
3cts emulsion paint to plastered walls.

ADD ←
2 B wall
    faced one side.

3·00
2·00

NOTE: it is good practice to make it clear when a measurement is to be added; especially after a DDT item.

**Figure 9.5** Typical dimension sheet entries.

**Keywords for measurement standards**

The following is a list of some keywords used in this chapter. Use the list to test your knowledge and, if necessary, consult the text to learn about the terms.

| | | |
|---|---|---|
| Measurement code | Preliminaries | Dimension paper |
| Practice manual | Measured work | Taking-off |
| General rules | Tabular format | Query sheets |
| Common arrangement | Supplementary rules | Take-off lists |

# *10   Measurement*

Measurement is the general process of analysing the working drawings of a building, or works already carried out, and using this information to produce details of the various amounts of labour and materials that will be needed, or that have already been used.

A basic process in measurement is the *taking-off* of quantities. Taking-off concerns the measurement of lengths, areas and volumes, a process sometimes referred to as *mensuration*. The purpose of mensuration is to quantify the amount of material and equipment required to complete a construction project.

## Mathematical techniques

The production of a Bill of Quantities may involve the following mathematical techniques:

- Ratios, proportions, percentages
- Trigonometric techniques
- Geometric techniques
- Lengths and perimeters
- Regular areas and volumes
- Irregular areas and volumes
- Use of formulas.

This chapter summarises some of the relevant mathematical techniques used in the processes of measurement. You should also separately learn and practise the mathematics required.

### Trigonometric techniques

Trigonometry is a branch of mathematics dealing with relations of sides and angles of triangles. Trigonometry is frequently used

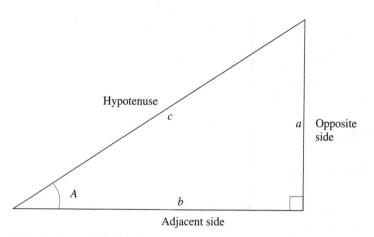

**Figure 10.1** Right-angled triangle.

to solve problems which arise in surveying practice.

The principal trigonometrical functions are as follows:

- Sine of an angle – often shortened to Sin
- Cosine of an angle – often shortened to Cos
- Tangent of an angle – often shortened to Tan.

These functions can usefully be described as combinations, or ratios, of the lengths of the sides of a right-angled triangle (Figure 10.1). For an angle $A$ in a right-angled triangle, labelled as shown in the figure, the trig functions can be defined as follows:

$$\text{Sin } A = \frac{\text{Side opposite angle } A}{\text{Hypotenuse side}} = \frac{a}{c}$$

$$\text{Cos } A = \frac{\text{Side adjacent to angle } A}{\text{Hypotenuse side}} = \frac{b}{c}$$

$$\text{Tan } A = \frac{\text{Side opposite angle } A}{\text{Side adjacent to angle } A} = \frac{a}{b}$$

A trig function must always be associated with a particular angle. For example, Sine $40° = 0.6428$ (rounded to 4 decimal places). This can be found from a simple scientific hand calculator.

The usefulness of the ratio formulas to surveying is that if two items are known (such as an angle and one length) then the third item (such as the other length) can be found by using the formula (see Figure 10.2).

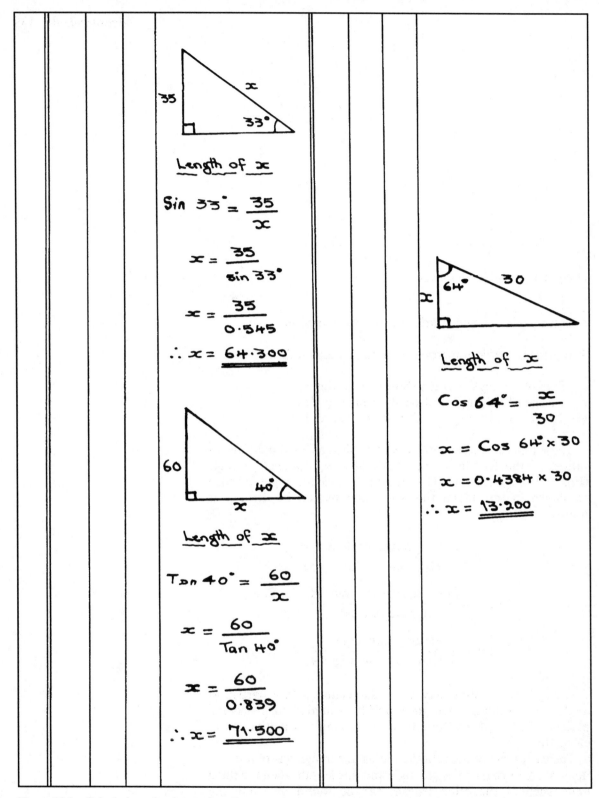

**Figure 10.2** Calculation of unknown lengths.

# Geometric areas

The exercise of measurement involves the Quantity Surveyor in measuring many different plan (flat) shapes. Some commonly-encountered types of areas are listed below, and formulas are given in the margin. Also refer to the diagrams (Figure 10.3) for shapes and labels.

- Square/Rectangle
- Triangle
- Circle
- Trapezium.

# Volumes

The Quantity Surveyor will encounter various 'mass' structures which have volumes. Some commonly-encountered types of volume are listed below, and formulas are given in the margin. Also refer to the diagrams for shapes and labels.

- Cube
- Pyramid
- Cylinder
- Frustum.

A 'frustum' is the portion of a cone or pyramid remaining when a horizontal slice removes the pointed top.

# Perimeters

During measurement, the most commonly-needed mathematical result is the length of the outer boundary, or *perimeter*, of a regular area such as those listed below. Some common perimeter formulas are given in the margin on page 142, and the diagrams (Figure 10.4) show shapes and labels.

- Square /Rectangle
- Triangle
- Circle
- Trapezium.

The examples on dimension sheets show typical entries for calculation of perimeters.

**Some area formulas**
$A$ = Area
$l$ = length
$w$ = width (breadth)
$h$ = height
$a$ = base length (trapezium)
$b$ = top length (trapezium)
$r$ = radius

For a rectangle:

$$A = l \times b$$

For a triangle:

$$A = \tfrac{1}{2} \times l \times h$$

For a circle:

$$A = \pi \times r^2$$

For a trapezium:

$$A = \tfrac{1}{2} \times h \times (a + b)$$

**Some volume formulas**
$V$ = volume
$l$ = length
$w$ = width
$A$ = area of base
$h$ = height
$r$ = radius

For a cube or box:

$$V = l \times w \times h$$

For a pyramid:

$$V = \tfrac{1}{3} \times A \times h$$

For a cylinder:

$$V = \pi \times r^2 \times h$$

For a sphere:

$$V = \tfrac{4}{3} \times \pi \times r^3$$

For a frustum:

$$V = \pi \times \left(\frac{r+R}{2}\right)^2 \times h$$

## Some perimeter formulas

$s$ = perimeter
$l$ = length
$w$ = width (breadth)
$h$ = height
$a$ = base length (trapezium)
$b$ = top length (trapezium)
$r$ = radius

For a rectangle:

$$s = 2l + 2w$$

For a triangle:

$$s = l + l + l$$

For a circle:

$$V = 2\pi \times r$$

## Mathematical operations

Addition
Subtraction
Multiplication
Division
Ratios
Formulas
Proportions
Percentages

## Mathematical operations

A surveyor needs to be competent in mathematical operations, such as those listed in the margin below.

The earlier examples have demonstrated the use of basic arithmetic calculations such as addition, subtraction, multiplication and division. Other operations include ratios, proportions and percentages.

### Ratios

A ratio indicates the relation between two or more numbers.

A mortar mix, for example, might have the following constituents:

300 kg of cement
900 kg of sand

By dividing the smaller number into the other number this relationship could be simplified to:

1 part of cement
3 parts of sand

This ratio is called 'one-to-three' and is written as $1:3$.

### Formula

A formula is a form of shorthand which uses symbols and figures and to express a mathematical relationship. The earlier expressions for areas and volumes are examples of formulas.

### Proportion

The ratios discussed above also indicate the relative proportions of the mortar mix. Two quantities are in *direct proportion* if an increase (or decrease) in one quantity is matched by an increase (or decrease) in the second quantity. Thus, if the amount of cement is doubled (from 300 kg to 600 kg) a proportional increase means that the amount of sand is also doubled (from 900 kg to 1800 kg).

### Percentage

A percentage is a proportionate or rate per hundred. Percentages are often used in the calculation of cost estimates. where it is sometimes necessary to add percentages for the risk element of the estimates.

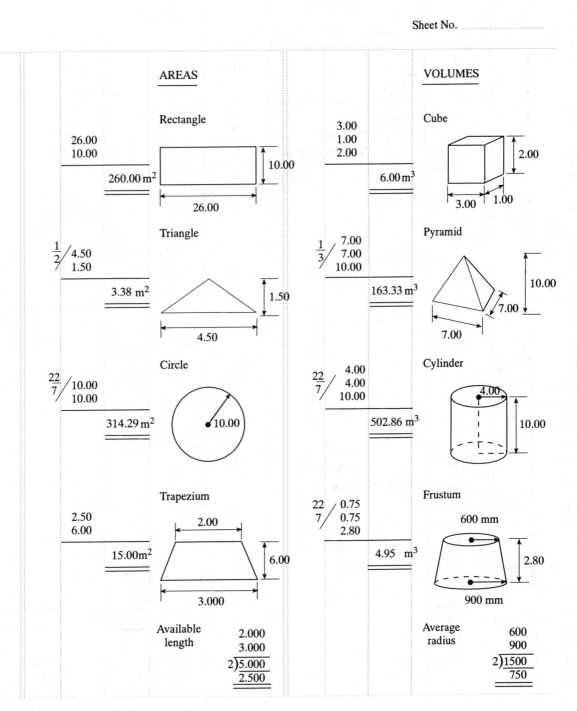

**Figure 10.3** Measurement of areas and volumes.

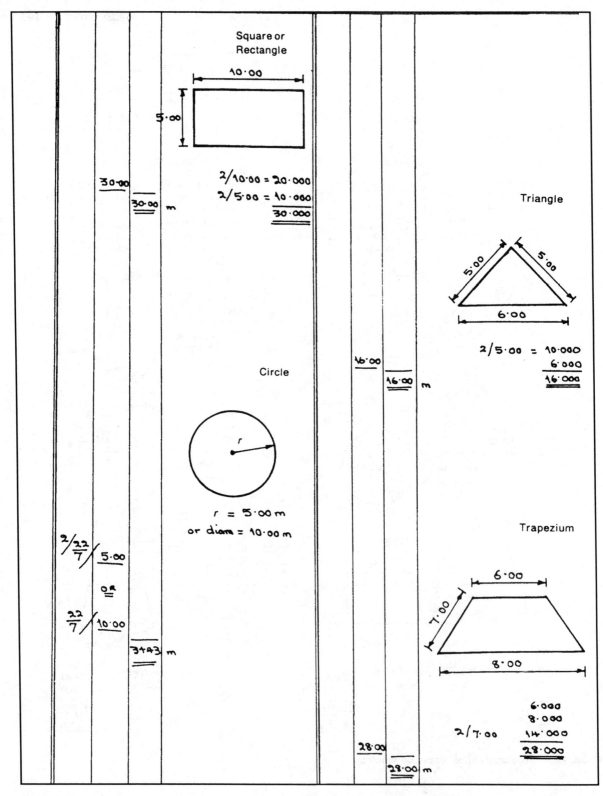

**Figure 10.4** Measurement of perimeters.

A percentage change, such as 5 per cent, is calculated by multiplying the original amount by the percentage figure and dividing by 100. A percentage can also be expressed as a decimal factor. For example, 5% is equivalent to a factor of 0.05.

Preliminary cost estimate

$$= £150,000$$

Add 5 per cent for risk

$$£150,000 \times \frac{5}{100} = £7,500$$

Total $= £157,500$

# Irregular areas and volumes

## Irregular areas

An irregular area is one where boundaries do not follow a definite pattern. The Quantity Surveyor is often asked to calculate areas of buildings, sites, roads and/or other features which are irregular in plan.

Various measurement techniques, such as the following, are used to find the area of irregular shapes:

- Mid-ordinate rule
- Trapezoidal rule
- Simpson's rule
- Bellmouth.

The first three methods (mid-ordinate rule, trapezoidal rule, Simpson's rule) use techniques which divide the area into strips of equal width. The accuracy increases as the number of strips increases.

An 'ordinate' is the length along one side of a strip, as shown in the diagrams (Figure 10.5). A mid-ordinate is the length of a strip measured at the middle point.

The parts of the formulas within brackets must be calculated first.

### Mid-ordinate rule

$$\text{Area} = \text{width} \times \text{sum of all mid-ordinates}$$

### Trapezoidal rule

$$\text{Area} = \text{width} \times [\tfrac{1}{2}(\text{sum of first and last ordinates}) + (\text{sum of remaining ordinates})]$$

### Simpson's rule

There must always be an even number of strips, which then give an odd number of ordinates.

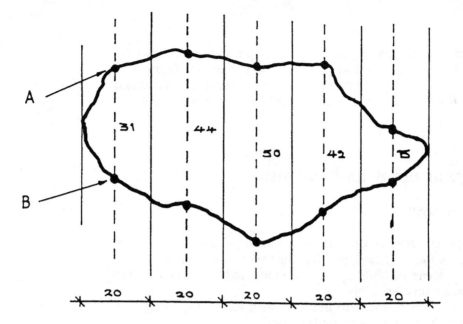

By Mid-ordinate Rule: $A = 20(31 + 44 + 50 + 42 + 15)$

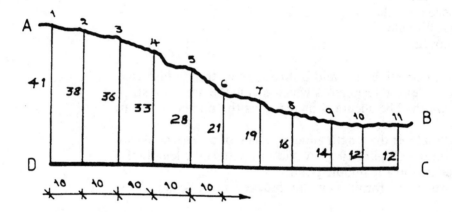

By Trapezoidal Rule: $A = 10[\frac{1}{2}(41 + 12) + (38 + 36 + 33 + 28 + 21 + 19 + 16 + 14 + 12)]$

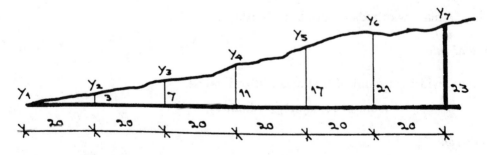

By Simpson's Rule: $A = \frac{20}{3}[(0 + 23) + 2(7 + 17) + 4(3 + 11 + 21)]$

**Figure 10.5** Examples of irregular areas.

Area = width/3 × [(sum of first and last ordinates)

+ 2(sum of remaining odd ordinates)

+ 4(sum of even ordinates)]

### Bellmouths

A bellmouth (Figure 10.6) is the term for the type of curved area which can occur at a road junction.

Area = Area of square − Area of quadrant (quarter circle)

$$= d \times d - \tfrac{1}{4}\pi r^2$$

which can be converted to

$$d^2 - \frac{\pi r^2}{4}$$

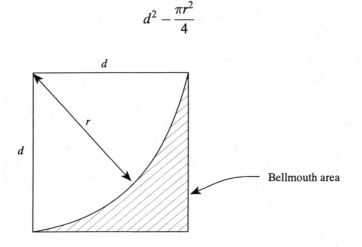

Area of bellmouth = area of the square *less* the area of the quadrant

$$= (d \times d) - \left(\pi r^2 \times \frac{90}{360}\right)$$

$$= d^2 - \frac{\pi r^2}{4}$$

**Figure 10.6**   Area of bellmouth.

## Irregular volumes

All the methods explained earlier for finding irregular areas may be adapted for finding irregular volumes, such as in an excavation.

### Prismoidal rule for calculating volumes

The *prismoidal formula* is a general formula by which the volume of any prism, pyramid or frustum of a pyramid may be found.

# Scheduling

**Scheduled components**
Joinery
Finishings
Furniture, ironmongery
Drainage
Sanitary appliances
Steel reinforcement
Mechanical and electrical
    fittings
Precast concrete components

A schedule is a tabular layout which lists items and gives the number or other details (see Figure 10.7). Scheduling is a convenient method of communicating design details about components which are repetitive and numerous. Schedules also assist with the measurement of certain works sections. A list of subject areas suitable for schedules is shown in the margin.

The use of schedules can give the following advantages:

- Time saved in taking-off
- Simplified checking of errors, omissions or description in tender documents
- Ease of access of information
- Logical presentation of information
- Ease of reference.

## SCHEDULE OF FINISHES

Job Name/Ref: _____   Block: _____   Sheet No. _____

| Room | Wall | Floor | Ceiling | Skirting | Additional information | Remarks |
|------|------|-------|---------|----------|------------------------|---------|
|      |      |       |         |          |                        |         |
|      |      |       |         |          |                        |         |
|      |      |       |         |          |                        |         |

## BAR REINFORCEMENT SCHEDULE

Job Name/Ref: _____   Block: _____   Sheet No. _____

| Bar mark | Type | Member | Diam. (mm) | No. of members | No. in each | Total No. | Length of each bar | Total length | Shape | Remarks |
|----------|------|--------|------------|----------------|-------------|-----------|--------------------|--------------|-------|---------|
|          |      |        |            |                |             |           |                    |              |       |         |
|          |      |        |            |                |             |           |                    |              |       |         |
|          |      |        |            |                |             |           |                    |              |       |         |

**Figure 10.7**   Example of schedules.

The most common applications of such schedules are in the following operations:

- Advance programming
- Checking deliveries
- Estimating
- Measurement of quantities
- Ordering materials
- Locating and carrying out work.

## Preparation of schedules

It is not possible to standardise the content and arrangement of schedules because of the diversity of the subjects. The preparation of a schedule is made easier by the following logical procedure for all subjects:

Step 1: Decide on the subject matter
Step 2  Design format of the schedule
Step 3: Identify individual components
Step 4: Identify components by reference to location.

It is important that information given in other tender documents is not duplicated in the schedule and vice versa. Nor should the information of one schedule duplicate or overlap that of another.

# Excavation

Quantity surveyors encounter a variety of earth shapes when measuring construction works and many of the mathematical techniques shown earlier in the chapter can be used. For example, the surveyor often needs to calculate the volume of soil from the excavation on a sloping site.

An example of a school playground site is shown in Figure 10.8. It is required to excavate down to a level of 18.000 m, including excavating topsoil to a depth of 150 mm. The average depth of excavation over the site is found by 'weighting' the depth at each point on the grid of levels, according to the number of areas affected by each level.

Step 1: Reference each point
Step 2: Draw up a table
Step 3: Plot cut/fill line
Step 4: Plot other SMM7 limit lines
Step 5: Weighting of areas
Step 6: Take-off quantities

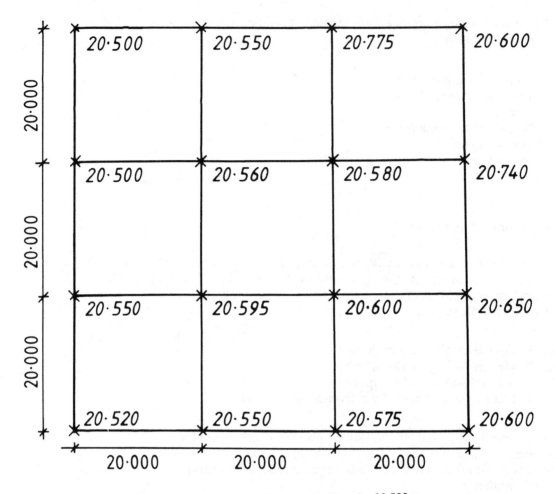

20 m grid. Overall plan size 60 × 60 m. Finished playground level = 20.500 m.
Playground construction consists of 150 mm (minimum) hardcore and 100 mm tarmacadam.

**Figure 10.8**   School playground – showing existing levels.

The technique involves taking the depths at the extreme corners of the area once; the intermediate points on the boundary twice; and all other intermediate points four times. The sum of the weighted depth is divided by the total number of weightings (number of squares by four) to give the average depth of the whole area.

When using this technique, it is important that the levels are spaced the same distance apart in both directions.

The surveyor will need to identify the various SMM7 *limit lines* for excavation (e.g. 0.25, 1.00) and filling (e.g. under 0.25 and over 0.25). It may also be necessary to identify a cut-and-fill line.

It is important that the Quantity Surveyor approaches the measurement of sloping sites in a logical order, as shown in the example (Figure 10.9).

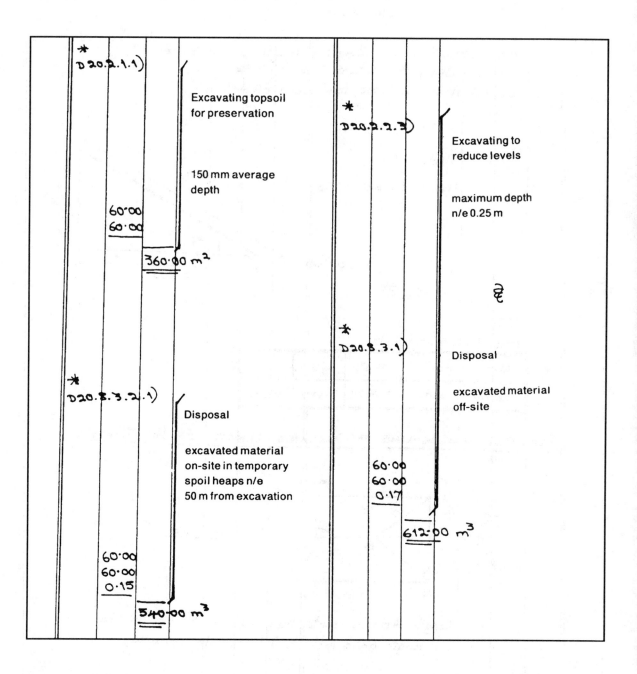

**Figure 10.9** Measurement of school playground.

# Pitched roofs

Mensuration principles are also used to measure the length of rafters, hip rafters, valley rafters and area of roof coverings (Figure 10.10).

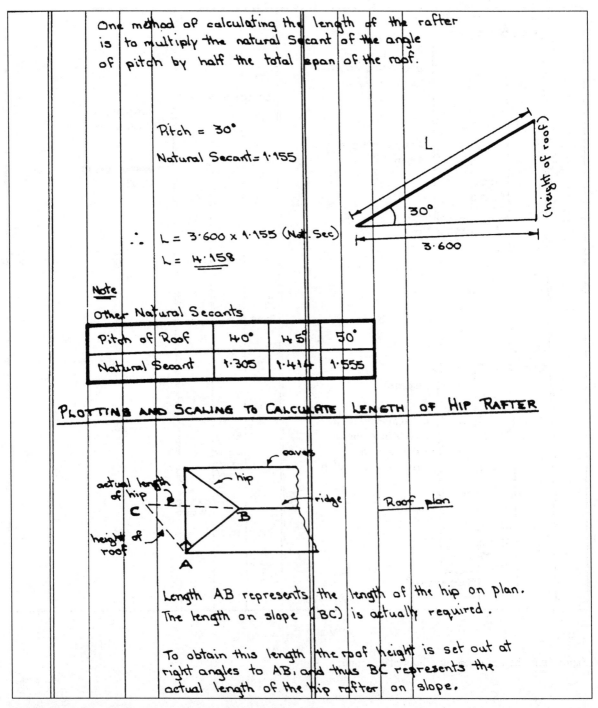

One method of calculating the length of the rafter is to multiply the natural Secant of the angle of pitch by half the total span of the roof.

Pitch = 30°

Natural Secant = 1·155

∴  L = 3·600 × 1·155 (Nat. Sec)

L = 4·158

Note

Other Natural Secants

| Pitch of Roof | 40° | 45° | 50° |
|---|---|---|---|
| Natural Secant | 1·305 | 1·414 | 1·555 |

PLOTTING AND SCALING TO CALCULATE LENGTH OF HIP RAFTER

Roof plan

Length AB represents the length of the hip on plan. The length on slope (BC) is actually required.

To obtain this length the roof height is set out at right angles to AB, and thus BC represents the actual length of the hip rafter on slope.

**Figure 10.10**   Calculations of common rafters.

The techniques available to the surveyor for measuring the length of sloping roof members include the following:

- Scaling from drawing
- Mathematical calculation
- Plotting and scaling.

# Perimeters of buildings

One of the most fundamental tasks in taking-off is to calculate the *girth* and *centreline* of a building or wall (Figure 10.11). These girth or centreline calculations start with the calculation of a perimeter using the dimensions given on the drawing.

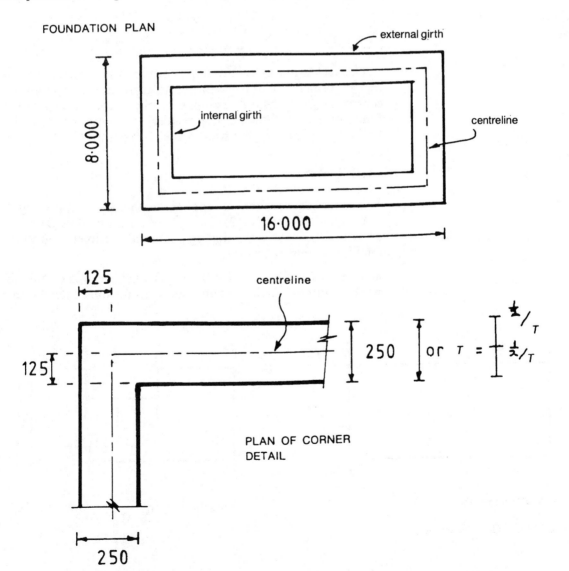

**Figure 10.11** Girth calculations.

**Corner adjustment**

Formula:

$$4 \times 2 \times \tfrac{1}{2} \times T$$

where

   4 = Number of corners

   2 = Number of adjustments at each corner

  $\tfrac{1}{2}$/T = One-half of wall thickness

   T = Wall thickness

---

### Example girth calculations

*External girth calculation*

$$\begin{array}{r} 8.00 \\ 16.00 \\ \hline \end{array}$$

$$2 \times 24.00 = 48.00$$

*Centreline calculation*

| External girth | 48.00 |
| --- | --- |
| *deduct* | |
| $4 \times 2 \times \tfrac{1}{2} \times 250$ | −1.00 |
| | 47.00 |

---

The technique of calculating centrelines and girths involves the 'movement' of this perimeter to provide other perimeters, centrelines or girths. For example, to obtain the centreline of the foundation plan shown in Figure 10.11, it is necessary to deduct half the thickness of the wall in each direction.

### Girth technique

The same 'girth technique' applies whether the shape is regular or irregular (Figure 10.12). The type of adjustment used depends on whether you are 'moving' external perimeter girths or internal perimeter girths.

- For *external* girth to centreline: Deduct formula calculation.
- For *internal* girth to centreline: Add formula calculation.

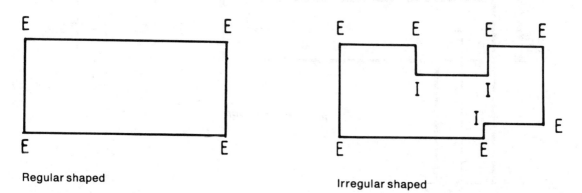

Regular shaped           Irregular shaped

**Figure 10.12**   Girth technique.

TRANSPOSING OF DIMENSIONS

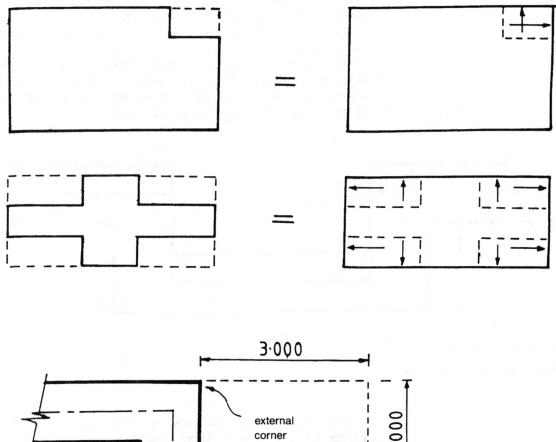

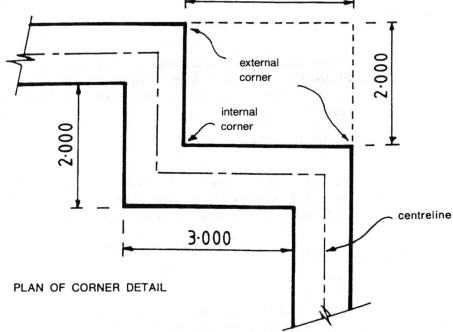

PLAN OF CORNER DETAIL

**Figure 10.13** Transposing of dimensions.

An additional factor in calculating girths or centrelines is whether a corner is classified as an external or an internal corner (Figures 10.12 and 10.13). The regular shape shown in the Figure 10.12 has four external angles, while the irregular shape has seven external angles and three internal angles. An inset or re-entrant of a corner would not affect the overall centreline calculations.

The situation is different if a re-entrant occurs somewhere along the wall other than at a corner, see Figure 10.14 for a re-entry wall. In this case, the surveyor must add twice the depth of the re-entrant to the girth or centreline.

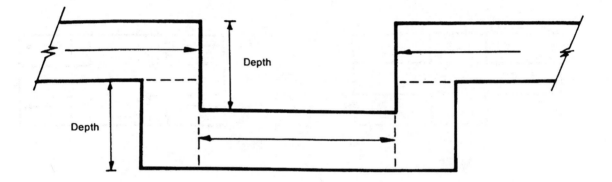

PLAN OF RE-ENTRANT

**Figure 10.14**   Re-entrant walls.

Examples of practical girth calculations are shown in Figure 10.15. In North America the technique of transposing of dimensions is known as the 'stretch-out-length' (SOL) concept.

### Stretched-out-length

The SOL can be defined as the length of the centreline of any strip of thickness $t$ which bounds the perimeter of a building foundation. The formula for determining the SOL is

$$SOL = P_o - 4t$$

where

SOL = Stretch-out-length
$P_o$ = Length of the outside perimeter
$t$ = Thickness or width of given strip.

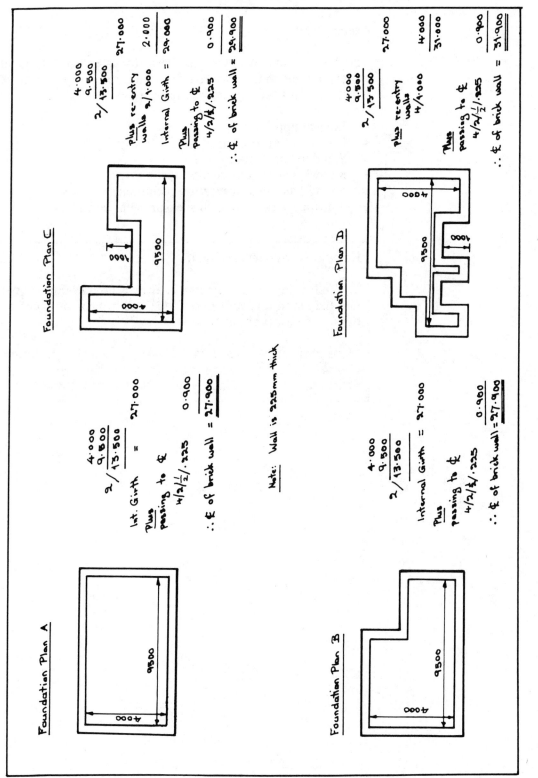

**Figure 10.15** Practical girth calculation.

157

### *Summary*

The aim of this chapter has been to establish the general understanding that measurement is basically a practical skill which involves the key skills listed below:

- Logical approach
- Consistent approach
- Standard presentation
- Knowledge of construction technology
- Sound basis in mensuration techniques
- Knowledge about use of measurement codes.

---

**Keywords for measurement**

The following is a list of some keywords used in this chapter. Use the list to test your knowledge and, if necessary, consult the text to learn about the terms.

| | | |
|---|---|---|
| Mensuration | Ratios | Girth calculations |
| Cosine | Proportions | Taking-off |
| Frustum | Scheduling | Percentages |

# 11 Bill of Quantities

A *Bill of Quantities* is a tender document, normally produced by a Quantity Surveyor, which contains general information on a construction project, together with quantities measured from drawings in accordance with a measurement code. The document will subsequently be priced by contractors and be used throughout the construction project for valuation and cost control purposes. The Bill of Quantities provides a wealth of information about a project, in addition to the detailed breakdown of the proposed work.

Bills of Quantities are firstly tender documents, and secondly contract documents. They are prepared in accordance with established measurement codes and conditions of contract provisions and should contain certain basic information and be presented in a recognised format to make them easy to use.

**Bill of Quantities content**
Preliminaries
Specification/Preamble
Measured works
Prime cost and provisional
    sums
Dayworks
General summary

## Functions of a Bill of Quantities

The overall purpose of a Bill of Quantities is to measure, in a systematic and standard manner, the work contained in the construction or alteration of a building. The aim is to obtain competitive tenders from contractors, subcontractors and suppliers.

### Advantages of a Bill of Quantities

Using a Bill of Quantities offers the following principal advantages:

- Each contractor prices an identical bill
- Contractors are relieved from preparing their own quantities.

- Contractor can concentrate on the tender estimate
- Risk of errors in measurement and pricing reduced because of preparation by professionals
- Labour-saving document for inviting competitive offers
- Locational identification of work
- Acts as a vehicle for valuing changes
- Assists contractor in the preparation of an estimate
- Assists in valuing work for stage payments
- Assists contractor in planning resources such as labour, plant and materials and organising project work
- Assists in preparing approximate estimates for future work
- Assists in preparation of final account at completion of project.

## Structure of a Bill of Quantities

A traditional Bill of Quantities (Figure 11.1) is divided into the following sections:

- Preliminaries
- Preambles clauses or specification
- Measured work
- Prime cost and provisional sums
- Dayworks
- General summary.

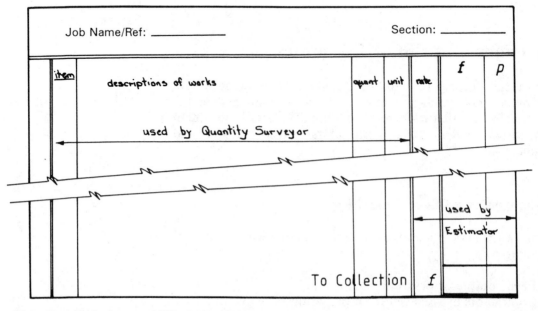

**Figure 11.1**   Typical blank page of Bill of Quantities.

# Content of a Bill of Quantities

## Preliminaries

The preliminaries section is normally the first section in a Bill of Quantities and its main purpose is to set out all the general liabilities and obligations of the contractor; it covers the entire contract.

This section outlines the general introductory information that a contractor needs to know when pricing. The section provides an opportunity to price work which is not closely related to measured quantities.

The preliminaries section normally contains the information listed in the margin and described below.

**Preliminaries section**
The project generally
Specific requirements of the
    project
The contract
Employer's requirements
Contractor's general cost
    items
Work by others, or work
    subject to instruction

---

### Items in preliminary section

*Project general nature*
General information relating to the client, Architect and Quantity Surveyor. Also gives an outline of the site and brief description of the work.

*Specific requirements*
List of any specific problems or employer's requirements that may affect contractor's pricing or method of working.

*Contract*
List of the conditions of the contract, together with details on any amendments to the standard printed conditions.

*Employer's requirements*
Identifies the principal requirements of the client in respect to quality control, security, safety, working restrictions and facilities to be provided for the client's benefit.

*Contractor's general cost items*
Identifies items that the contractor must provide for the proper running of the site, such as management, site accommodation, services, facilities, plant and temporary works.

*Work by others or subject to instructions*
Describes work or materials arranged directly by the client or at its instruction, such as work by statutory bodies, nominated subcontractors and suppliers, and provisional work including daywork.

## Preambles section

The *preamble* section contains details of the specification in terms of the work, the workmanship, and the materials; together with any further information that may qualify the scope and interpretation of measured work item descriptions. The estimator needs to know about preamble clauses as they affect the rates that are inserted against billed items of measured work.

Preambles clauses allow the description of the measured work items to be reduced significantly in length and assist with the pricing. The contractor does not price the preamble section directly, but will be influenced by the information contained in that section.

The preambles section is found in the Bill of Quantities in two alternative positions:

- After the preliminaries section
- At the start of each work section.

---

### Example of trade preamble

*Excavation and filling*

A   The rates included for excavation are to include for excavation in all types of ground.

B   In the event of the contractor excavating below the given levels, then the contractor will be required to fill extra depth with lean-mix concrete (1 : 12) at its own expense.

C   The contractor shall divert as required and protect existing services encountered during building work.

D   The 'pre-contract water level' was established as being 3.20 m below existing ground level on 28 July 1997.

---

## Specification

The *specification* is sometimes referred to as the preambles to trades sections and consists of general descriptions of materials and workmanship requirements. The specification may be presented in a number of ways, depending on the current preferences of architectural practices.

The specification is normally divided into trades and considers the following two elements:

- Materials
- Workmanship.

A specification may supplement a Bill of Quantities, or it may form a full specification in its own right. If it is a full specification, it will also include a third element: a description of the actual work to be executed.

---

**Example of specification**

*Specification*

*Brickwork and blockwork*

*Materials*

A  *Foundation bricks* 101
102.5 mm brickwork in cavity construction in calcium silicate bricks class 3.

B  *Damp-proof courses* 102
Bitumen as BS 743, Type 5A, hessian base and joints well lapped.

C  Ties 103
Copper butterfly wire wall ties to comply with BS 1243, Type B, and 150 mm long for 50 cavity.

D  Workmanship
All bricks to be neatly stacked on site and not tipped and should be protected from the elements.

E  All mortar to be mixed on site in mechanical mixers adding only sufficient water to give correct consistency.

---

## Measured work

This *measured work* section represents the bulk of the document and presents measured quantities in a standard format which are measured in compliance with the rules of a particular standard method of measurement.

Each item of work is listed separately and consists of a description, unit of measurement and quantity with a column to enable the contractor to insert the unit rate. The unit rate multiplied by the measured quantity will provide the contractor's price for that item of work. The items are often grouped together into work sections corresponding to the order set out in the relevant measurement code.

Each part of the work section is priced and totalled and taken to the general summary at the rear of the Bill. An example of a measured work extract is shown in Figure 11.2.

## Prime cost and provisional sums

This prime cost and provisional sums section of the Bill contains details of work which is subject to instruction by someone else, normally the Architect. The following is an account of the sums inserted in this section.

### Provisional sums

*Provisional sums* are included in a Bill of Quantities to allow for work or for costs which cannot be entirely foreseen, defined or detailed at the time that the tendering documents are issued.

Rather than delay production of the Bill of Quantities, an approximate sum of money is included to cover the likely cost of such work. Provisional sums can also be used to cover the cost of works undertaken by the main contractor or a specialist subcontractor.

Under the particular rules of the SMM7 measurement code these sums are now divided into subheadings entitled 'Defined' or 'Undefined'.

Where the sum is defined the contractor must allow for the associated programme and time implications within the tender.

Where the sum is undefined, the contractor is entitled to receive an additional allowance to the programme, together with payment for any associated additional preliminaries costs, for work ordered against such undefined work.

An example of a provisional sum is shown in Figure 11.2.

### Prime cost sums

It is common for the Architect to appoint specialist subcontractors or suppliers of building materials and goods. Such companies are known as nominated subcontractors or nominated suppliers. An approximate sum of money is allocated within the Bill of Quantities to cover each of these specialists.

Job Name/Ref: _____          Section: _____

| | MEASURED WORK | | | £ | p |
|---|---|---|---|---|---|
| | **Superstructure** | | | | |
| | **F10 Brick/Block Walling** | | | | |
| | **Common brickwork, as Spec. clause 302** | | | | |
| a | Wall<br>Half brick thick;<br>     facework one side | 56 | m² | | |
| b | One brick thick;<br>     curved on plan to a 1350 mm<br>     mean radius;<br>         facework both sides | 34 | m² | | |
| | **PROVISIONAL SUMS** | | | | |
| c | Include the defined Provisional Sum of £500.00<br>for the provision of an additional 25 m of<br>25 × 150 mm shelving | Item | | 500 | 00 |
| | **PRIME COST SUMS** | | | | |
| d | Provide the PC Sum of £750.00 for automatic<br>entry doors to be delivered to site by a supplier<br>nominated by the Architect | | | 750 | 00 |
| e | Allow for profit | Item | | | |
| | | To Collection  £ | | | |

**Figure 11.2** Example of measured work, provisional sums and prime cost sums.

In practice the term 'prime cost' is often shortened to PC sum when used in the everyday context of building operations. Prime cost sums may also be used to cover work by local authorities or public undertakings such as relevant water, gas and electricity companies.

The sum inserted may take one of two forms:

- Lump sum items: to be spent as directed, or
- Measured items: the descriptions containing a unit price which the estimator must allow for in the tender build-up.

An example of a typical entry in the Bill of Quantities is shown in Figure 11.2.

### Dayworks

The term *dayworks* refers to a method of payment to a contractor for additional work ordered by an Architect during the contract that cannot be properly measured and valued using original or adjusted bill rates (see Figure 11.3).

The Bill contains notional monies for labour, materials and plant. Under each of the three components there is a provision for the contractor to insert a percentage addition or deduction. The percentage inserted must cover all monies required by the contractor over and above the prime (net) cost to the contractor. The prime cost for each component is in accordance with the RICS Definition of Dayworks carried out under a building contract.

Explanation of the prime cost allowance for labour, plant and materials are shown in the table.

---

**Prime costs**

*Prime cost of labour*

Prime cost is made up of the sum of the operatives' guaranteed wages plus the following: other agreed payments; extra payments for skill etc.; Public Holidays payments; employer's National Insurance; annual holiday and death benefit scheme contributions; statutory contributions made by the employer.

*Prime cost of material*

The cost of materials delivered to site, less all discounts other than cash discount.

*(continued)*

**Prime costs** (*continued*)

*Prime cost of plant*

This prime cost is normally provided by the contract. A common arrangement is that net plant costs are valued in accordance with the current schedule of basic plant charge issued by the Royal Institution of Chartered Surveyors. The schedule is intended to serve as a base index and therefore the prices do not reflect current market prices.

## Contingencies

It is common practice for a sum to be inserted in the Bill of Quantities by an Architect to be spent on unexpected items. This sum gives the Architect a sum of money to cover minor alterations to the project without the need to approach the client for additional funds. The contractor is not required to make any percentage adjustment to the sum.

# Bill of Quantities types

A Bill of Quantities can be presented in a variety of formats, described below.

**Bill of Quantity types**
Traditional
Elemental
Sectionalised
Operational
Activity

## Traditional

The traditional format (Figure 11.4) gives a Bill of Quantities measured and divided into trade or work sections in accordance with the measurement code.

## Elemental

The elemental format (Figure 11.5) gives a Bill of Quantities measured and divided into elements in accordance with the standard form of cost analysis issued by the Royal Institution of Chartered Surveyors Building Cost Information Services.

An 'element' is that part of a building which always performs the same function, irrespective of the type of building. Example of elements include internal partitions, internal doors and roofs.

Job Name/Ref: _____     Section: _____

| | | | £ | p |
|---|---|---|---|---|
| | **DAYWORKS SCHEDULE** | | | |
| a | Include the undefined provisional sum of £1,800.00 for labour | | 1,800 | 00 |
| b | ADD for overheads and profit | % | | |
| c | Include the undefined provisional sum of £1000.00 for materials | | 1,000 | 00 |
| d | ADD for overheads and profit | % | | |
| e | Include the undefined provisional sum of £500.00 for plant | | 500 | 00 |
| f | ADD for overheads and profit | % | | |
| | The foregoing Provisional Sums apply to Dayworks ordered by the Architect prior to the commencement of the Defects Liability Period. Should any Dayworks be necessary after the Defects Liability Period it is proposed that the definition shall also apply to such work, and the Contractor is invited to insert an appropriate percentage addition for Overheads and Profit to the following: | | | |
| g | Labour | % | | |
| h | Materials | % | | |
| j | Plant | % | | |
| | To Collection   £ | | | |

**Figure 11.3**  Example of dayworks schedule.

| | | | £ | p |
|---|---|---|---|---|
| | Job Name/Ref: _____ | | Section: _____ | |

**TRADITIONAL BILL FORMAT**

**F10 BRICK/BLOCK WALLING**

**Celcon partition blocks, as Spec. clause no. E210 in gauged mortar (1:1:6), and pointed with a neat flush joint as the work proceeds**

Walls

| | | | | £ | p |
|---|---|---|---|---|---|
| a | 100 mm thick | 100 | m² | | |
| b | 100 mm thick<br> facework one side | 80 | m² | | |
| c | 100 mm thick<br> curved on plan to 0.90 m mean radius | 21 | m² | | |
| d | 200 mm thick | 15 | m² | | |

**M20 PLASTERED COATINGS**

**14 mm thick plaster to BS 1191 Part 2, comprising 12 mm thick undercoat, 2 mm thick finishing coat trowelled smooth**

Walls

| | | | | £ | p |
|---|---|---|---|---|---|
| e | over 300 mm wide<br> to blockwork | 100 | m² | | |
| f | over 300 mm wide<br> to brickwork | 120 | m² | | |

**Note**: Extracts from **two** work sections of SMM 7 are shown. The order of the work sections and the items within them generally follow this layout of the SMM

To Collection £

**Figure 11.4** Example of traditional bill format.

|   |   | £ | p |
|---|---|---|---|

**ELEMENTAL BILL FORMAT**

**F10 BRICK/BLOCK WALLING**

**Celcon partition blocks, as Spec. clause no. E210 in gauged mortar (1:1:6), and pointed with a neat flush joint as the work proceeds**

Walls

| | | | |
|---|---|---|---|
| a | 100 mm thick | 100 | m² |
| b | 100 mm thick facework one side | 80 | m² |
| c | 200 mm thick curved on plan to 1500 m mean radius | 20 20 | m² m² |

**WALL FINISHES (Element 3A)**

**M20 PLASTERED COATINGS**

**14 mm thick plaster to BS 1191 Part 2, comprising 12 mm thick undercoat, and 2 mm thick finishing coat trowelled smooth**

Walls

| | | | |
|---|---|---|---|
| d | over 300 mm wide to concrete | 90 | m² |

**M60 PAINTING/CLEAR FINISHING**

**Painting, plaster: prepare and apply one mist and two full coats of emulsion paint**

General surfaces

| | | | |
|---|---|---|---|
| e | over 300 mm wide | 90 | m² |

To Collection    £

**Figure 11.5**  Example of elemental bill format. *Note*: Extracts from two elements of the RICS Building Cost Information Service (BCIS) Standard Form of Cost Analysis are shown. Elemental references are quoted. The order of the section follows the layout of the SMM.

## Sectionalised

The sectionalised format gives a Bill of Quantities measured and divided into trades, but with each trade sectionalised into elements. This format was designed to have the advantages of both the traditional and elemental formats.

## Operational

The operational format gives a less-common type of Bill of Quantities which divides the work up into actual site operations. An operation is the amount of work which can be produced by a gang of men (between break points in the programme), independently of work by another gang, or without interruption, by the other gang.

## Activity

The activity format gives an operational Bill of Quantities which is measured in accordance with the standard method of measurement, but in which the items are sorted in accordance with a previously-agreed programme of activities. This form is also less common than the first three.

# Production of a Bill of Quantities

The methods used to prepare Bills of Quantities vary between different quantity surveying practices or firms, but usually take one of the forms listed in the margin. Some practices may use a combination of more than one system and make use of new ideas. However, the most widely used method is the *Group method* (or London method).

**Bill production methods**
The Group method
  (or London method)
The Northern method
  (or trade-by-trade method)
Cut and shuffle
Billing direct
Computer application
Shorter Bill of Quantities

## Group method

In this method the Bill is grouped into a special order for taking-off purposes and regrouped into the required Bill format before the Bill is printed. The order in which the taking-off is prepared is designed to assist in the speed and accuracy, and often bears little relation to the order in which items will finally appear in the Bill. This procedure has also been termed *traditional abstracting and billing*.

The processes involved in producing a completed Bill of Quantities require the following stages:

- Taking off
- Squaring and checking
- Abstracting and checking
- Reducing and checking
- Writing draft Bill of Quantities
- Editing draft Bill of Quantities and checking
- Proof reading the Bill of Quantities and checking.

The stages after the taking-off are known as *working-up*. A brief explanation of these stages is given.

---

**Stages for production of Bill of Quantities**

*Stage 1: Taking-off*

Dimensions are set down on dimension sheets in the normal manner.

The taking-off procedure is not affected by the method used in producing the Bill of Quantities.

*Stage 2: Squaring and checking*

Dimensions are squared out (totalled) and then checked for accuracy.

*Stage 3: Abstracting*

Similar descriptions are collected or collated, together with their quantities, on specially-ruled paper known as abstract or analysis paper. The descriptions are written in the final form and order that they will appear in the printed Bill of Quantities.

Once complete, the abstracting should be checked by another person and any errors returned to the original abstracter.

*Stage 4: Reducing and checking*

The *add* and *deduct* quantities are squared and reduced to a net quantity, which is then rounded up or down to the nearest whole unit. This net figure is then inserted adjacent to the units in the abstracting sheets.

Once the reducing is complete, it should be checked by another person, and any errors are returned to the original reducer.

*(continued)*

---

---

**Stages for production of Bill of Quantities** (*continued*)

*Stage 5: Writing draft Bill of Quantities*

The draft Bill of Quantities is written in its final form, ready for printing. The draft should then be checked for accuracy by another person.

*Stage 6: Editing draft Bill of Quantities*

The draft Bill of Quantities is edited by carrying out all the normal spot checks on individual items and sections.

*Stage 7: Proof reading Bill of Quantities*

The documents are proof read in full by an appointed person, normally a Senior Quantity Surveyor or other partner in the organisation.

After being word-processed, bound and packaging, the documents should be finally checked before being sent out to tender.

---

## The Northern method

In this method the quantities are taken-off in the order in which they will subsequently appear in the Bill. With this technique the sorting of items into Bill order is minimal, or unnecessary.

The technique allows for each trade to be prepared for printing while the next trade is being taken-off. The method is also known as the *trade-by-trade system*.

## Cut and shuffle

In this method quantities are taken off in exactly the same manner as described earlier, except that each description with its dimensions is written on separate slips of paper. When the measurement is complete the slips are sorted into Bill order and slips containing the same descriptions are brought together and totalled on each slip.

This cut and shuffle system eliminates the need to prepare the abstract and write the draft bill and therefore saves time and much of the labour in the working-up stage. The person taking-off must, however, aim to write all descriptions in their final form as they are, in fact, writing a draft bill.

### Billing direct

This method of producing a Bill of Quantities is similar to the traditional method, except that this method eliminates the abstracting process. The draft Bill is written *directly* from the original take-off.

### Computer applications

Computer technology can be used to produce a Bill of Quantities in any of the formats discussed earlier and computer methods are becoming standard practice. A variety of software packages is available from leading software houses and the technology in this area moves rapidly.

The technology of telecommunications and electronic data interchange also makes it possible to pass documents rapidly. So a Bill of Quantities, for example, can be sent to interested parties, such as contractors, subcontractors, the Architect, engineers and other consultants.

### Shorter Bill of Quantities

A shorter Bill of Quantities has been developed to meet changing demands in the industry for standard but concise Bills of Quantities. The aim of the shorter Bill of Quantities is to achieve a compromise between the over-simplification of brevity and the excessive complexity of a full Bill of Quantities.

Shorter Bills of Quantities have the following main advantages:

- Less complex than SMM6 and SMM7
- Up to 50% fewer items to measure
- Costs less to produce
- Stand-alone system requiring no reference to other documents
- Readily available computer software
- Follows conventional trade format
- Cost control easier since secondary items are less significant
- Simpler extraction of elemental cost data.

# Coordinated Project Information (CPI)

The Coordinating Committee for Project Information (CCPI) was established by the RIBA, RICS, BEC and ACE, with the object of bringing about a general improvement and standardisation in the documents used for the procurement and construction of buildings.

The main documents considered were drawings, specifications and Bills of Quantities. The committee suggested improvements to the technical content of tender documents and examined the effectiveness of the coordination between them.

The conventions for coordination were designed in the main for Architects, engineers, surveyors and others who are involved in the production of documentation.

The coordination conventions topics are listed in the margin. The core of the conventions establishes a common arrangement of work sections abbreviated as CAWS.

**Coordinated conventions**
Common work sections
Production drawings
Projection specification:
   National building
     specification
   National engineering
     specification
Bill of Quantities
SMM7 standard
   descriptions:
   Bill of Quantities

---

### Keywords for Bill of Quantities

The following is a list of some keywords used in this chapter. Use the list to test your knowledge and, if necessary, consult the text to learn about the terms.

| | | |
|---|---|---|
| Bill of Quantities | Daywork | Northern method |
| Preliminaries | Contingencies | Cut and shuffle |
| Preambles | Provisional sums | Billing direct |
| Specification | Prime cost sums | Working-up |
| Measured work | London method | Abstracting |

# 12 *Pricing*

The process of pricing, or estimating as it is commonly called, is carried out by an estimator. An estimator works for a contractor and uses the completed Bill of Quantities after it has been sent to selected contractors for pricing. The main sections of the Bill of Quantities which are priced by the contractors are the preliminaries section and the measured works section that are described in the previous chapters.

This chapter concentrates on *unit rate estimating* which is the process where the estimator calculates a unit rate for carrying out a specific item of building work. This unit rate is calculated in terms of cost per linear, square or cubic metre; or number of items, or tonnes in the case of reinforcement or structural steelwork. The price is then inserted in the rate column of the Bill of Quantities and 'timesed' (multiplied) by the quantity of the work to arrive at the total cost for that item of building work.

A sample of a typical price entry to a Bill of Quantities item is shown in Figure 12.1.

An estimator needs to have the following skills:

- Ability to read and quickly understand technical documents and drawings
- Good organisation and communication skills
- Good numerical skills
- Good understanding of quantity surveying measurement skills
- Creativity and imagination
- Clear understanding of the provisions of construction contracts.

The items in a typical basic unit rate for a square metre of brickwork are shown in the margin.

**Unit rate for brickwork**
(Square metre)
Cost of bricks
Waste of bricks
Mortar
Waste of mortar
Mortar mixer
Scaffolding (if required)
Cost of bricklayers
Cost of attending labour

| Project Name/Ref.: | Section |
|---|---|

**F10 Brick/Block Walling**

**Common bricks, to BS 3921, in cement mortar (1:3) and pointed with a neat flush point as the work proceeds**

Walls

| | | | | | | |
|---|---|---|---|---|---|---|
| a | 105 mm thick | 112 | m² | 49.75 | 5572 | 00 |
| b | 215 mm thick<br>    fair face both sides | 78 | m² | 77.50 | 6045 | 00 |
| c | 215 mm thick<br>    curved on plan to a 0.90 m mean radius | 45 | m² | 83.50 | 3757 | 50 |
| d | 330 mm thick | 15 | m² | 109.75 | 1646 | 25 |

**Figure 12.1** Typical price entry to Bill of Quantities.

# Composition of unit rates

The estimator calculates the unit rate by breaking down the rate into the following components:

- Labour
- Materials
- Plant
- Overheads
- Profit.

The Bill items are broken down into their component parts which are each priced separately. The labour and plant elements are normally compiled from a library of output standards, but standard published rates are built up from first principles. The material elements are built up from supplier quotations with adjustment for wastage, conversion factors and discount.

**Unit rate components**
Labour
Materials
Plant
Overheads
Profit

The pricing procedure is summarised in the margin and the components of a unit rate are described in the following sections.

## Labour

The unit price for labour is made up of basic cost of labour, which may vary in each district, *plus* the statutory costs payable automatically by any employer, and any additional payments required by the current Working Rule Agreement (WRA).

The unit price for labour initially involves building up an *all-in hourly rate* which will include the following components:

- Basic rate
- Non-productive overtime
- Inclement weather
- Sick pay
- Incentive payments
- Travelling allowance
- Trade supervision
- Working Rule Agreement specific payments
- National Insurance contributions
- Training levy (to CITB or similar)
- Annual and public holiday pay
- Redundancy or severance pay
- Employer's liability insurance
- Third party insurance.

An example of an all-in-hourly rate for labour is included in the examples.

### Pricing Policy

Gather sources of information
⇓
Calculate resource costs of:
Labour–Materials–Plant
⇓
Calculate
net unit      ⇒ Synthesis of
rates            unit rates
⇓
Calculate      Add allowance
gross unit  ⇒ for overheads
rates            and profits
⇓
Insert unit
rates in Bill
of Quantities

## Example calculation of an all-in hourly rate

|  |  | £    p |
| --- | --- | ---: |
| (a) | Basic rate and guaranteed minimum bonus | 9,810.00 |
| (b) | Inclement weather | 240.00 |
| (c) | Non-productive overtime | 445.00 |
| (d) | Sick pay | 75.00 |
| (e) | Trade supervision | 500.00 |
| (f) | Working rule allowance (extra payments) | 200.00 |
|  |  | 11,270.00 |
| (g) | Training levy (e.g. @ 0.25% on £11,270.00) | 28.18 |
| (h) | National Insurance (e.g. 12.5% on £11,270.00) | 1,408.75 |
| (i) | Holiday (i) annual | 1,040.00 |
|  |      (ii) paid public holidays | 305.00 |
|  |  | 14,051.93 |

(j)   Redundancy and Severance payments
          (e.g. 2.0% on £14,051.93)                                 281.04
                                                             _____
                                                           14,332.97
(k)   Employers' liability (e.g. 2.5% on £14,332.97)         358.33
                                                             _____

      *Annual cost per operative*                          14,691.30
                                                             =======

**Hourly rate**
Number of productive working hours worked in one year (after deduction for inclement weather) is 1873 hours

$$\text{Rate per hour} = \frac{£14{,}691.30}{1873} = £7.85/\textbf{hour}$$

The estimator must use skills of judgement and experience to calculate the labour constants. This process is assisted by data published in price books and estimating supplements, and by the estimator's information from previous schemes. A labour constant can be described as a constant time allowance to be included in a unit rate build-up to cover a particular item of work. Typical labour constants are shown in the examples.

The object of the exercise is to use this hourly rate and convert this rate into a measured unit rate. This is achieved by multiplying the all-in hourly rate by the selected labour constant, as shown in the examples.

## Typical *labour constants* for brickwork

| | |
|---|---|
| Forming cavities | $0.03\,\text{hr/m}^2$ |
| Closing cavities vertically | $1.00\,\text{hr/m}^2$ |
| Closing cavities horizontally | $1.00\,\text{hr/no.}$ |
| Building-in precast lintels (small) | $0.15\,\text{hr/no.}$ |
| Building-in precast lintels (large) | $0.30\,\text{hr/m}^2$ |

## Typical calculation of *unit rates*

| Item | Description | Hourly rate | Labour constant | Labour unit rate £ p |
|------|-------------|-------------|-----------------|----------------------|
| a | Forming cavities | 7.85 | $0.03\,\text{hr/m}^2$ | 0.24 |
| b | Closing cavities vertically | 7.85 | $1.00\,\text{hr/m}^2$ | 7.85 |
| c | Bonding in precast lintels (small) | 7.85 | $0.15\,\text{hr/no.}$ | 1.18 |

## Materials

The material costs for a project are usually calculated by quotations from suppliers for that project. The factors to consider in pricing the material element are summarised below.

---

### Factors for material pricing

*Work measured net*

Under the SMM contract work is measured 'net', as fixed in position, with no allowance for waste, for laps on sheet materials, or for bulking, consolidation or shrinkage of materials. The contractor therefore needs more than the measured quantities to complete the work and this extra cost must be added to the unit rate.

*Delivery and handling cost*

An allowance is normally added to the labour cost to cover costs of delivering and handling materials.

*Allowance for discounts*

A discount refers to a 'deduction from the usual price' given by a supplier. The estimator should be familiar with the discounts obtained by his company, ensure that discounts are secured, and make appropriate adjustments to the material prices.

*Conversion factors*

The preparation of unit rates requires the use of many conversion calculations in order to bring all calculations to a common denominator.

---

*Factors for material unit rate*

A typical breakdown of a material unit rate will include allowance for the following factors:

- Basic cost of materials purchased
- Allowance for wastage
- Allowance for laps, e.g. damp-proof course
- Allowance for delivery, unloading and storage
- Allowance for packaging

- Conversion factor for mixing, bulking, or consolidation
- Allowance for trade discount
- Allowance for method of measurement
- Cost of any secondary fixing materials.

## Typical material unit rate

| Ref. | Description | Quantity | Unit | Rate £ p | Unit rate £ p |
|---|---|---|---|---|---|
| Nat | For 1 m³ of mortar | | | | |
| | Cement | 0.26 | t | 85.00 | 22.10 |
| | Lime | 0.13 | t | 90.00 | 11.70 |
| | Sand | 1.71 | t | 9.00 | 15.39 |
| | Rate for 1:1:6 mortar | 1 | m³ | | 49.19 |

## Plant

The unit rate for plant included by an estimator will depend on whether the contractor owns its own plant, or decides to hire the plant for that particular project. The ownership of plant will be a critical factor to consider when building up a unit rate.

It is becoming less common for contractors to own their own plant and more likely that they will hire plant for particular contracts. A contractor needs to decide whether it is more economical to buy plant or to hire plant. Each contract should be examined on its own merits.

**Ownership advantages**
Convenience and flexibility
Plant can be purchased for
    particular purposes
Tax advantages

**Ownership disadvantages**
Ties up capital
Cost of maintenance and
    transport
Financing costs
Cost of idle plant time

---

*Factors for price of plant*

The unit price for the plant element of the all-in rate depends on the following factors:

- Size and nature of the site
- Size and nature of the structure
- Ownership of the plant
- Continuity of plant usage
- Amount of idle time.

---

If a contractor purchases an item of plant, the cost of using the plant should be calculated, bearing in mind the following considerations:

- Purchase price of plant and its expected economic life
- Return on capital
- Maintenance and repair
- Running costs (e.g. petrol, oil and diesel)
- Operatives' wages
- Transport costs to and from site
- Insurance and road licence tax.

If a contractor decides to hire plant, then basic hire charges can be obtained from an individual plant hire firm, or from one of the schedules of plant hire charges published by the Building Employers Confederation or by professional institutions. The estimator should then add the following costs to the basic hire charges, provided they are not included in the basic rates:

- Running costs
- Operation wages
- Transport costs.

An example of a unit rate build-up is shown in the example below.

### Typical calculation of *all-in plant rate*

Hourly rate for cost of operating a $5/3\frac{1}{2}$ mixer on site when weekly charge is £35.00

*Assumptions*
1. Week worked is 5 days = 39 hours
2. Standing time of 1 hour per day allowed for stoppages, bad weather etc.
3. Half an hour per day allowed for maintenance time

Total number of productive hours per week

$$= 39 - 5 \times (1 + \tfrac{1}{2})$$
$$= 39 - 7.5$$
$$= 31.5$$

| *Effective hourly hire rate of mixer* | | £ p |
|---|---|---|
| Hire cost: | £35.00/31.5 hours | 1.12 |
| Consumable costs: | 3 litres of fuel @ 50p per litre | 1.50 |
| | for oil, grease etc. | 0.50 |
| Operator costs: | All-in rate | 7.85 |

| Maintenance costs: | $\frac{1}{2}$ hour per day | |
|---|---|---|
| | = 2.5 hr per week @ £7.85 | |
| | = £19.60 per week ÷ 31.5 hr | 0.63 |

| | |
|---|---|
| Total cost hourly operation (including operator) | £11.60 |
| or | |
| Total cost hourly operation (excluding operator) | £3.75 |

## Overheads

Overheads are the cost to the contractor of running the business. The list below identifies some typical costs of running a business:

● Office accommodation
● Office lighting and heating
● Telephone and postal charges
● Salaries
● Company vehicles
● Rates or rent
● Bank charges
● Electricity and fuel
● Maintenance
● Insurances
● Company pension schemes
● Stationery and incidentals.

A contractor should endeavour to recover the costs of overheads from each project for which the contractor tenders. The correct percentage to be added for these *establishment charges* can be obtained by calculating the relationship between the annual turnover and the actual cost of the establishment charges. This relationship is obtained by calculating the cost of overhead expenses for the year and predicting the likely value of the work obtained. An example of a typical overhead percentage is shown in the illustration below.

## Typical *overhead percentage* calculation

|  | £      p |
|---|---|
| Gross turnover | 2,000,000.00 |
| *Less* profit | 100,000.00 |
|  | 1,900,000.00 |
| *Less* general overheads | 250,000.00 |
| Net annual cost | 1,650,000.00 |

General overheads as a percentage of net cost (turnover)

$$\frac{250,000.00}{1,650,000.00} \times 100 = 16 \text{ per cent}$$

## Profit

Contractors expect to earn a profit when they carry out work. The building industry covers a wide variety of work and profit levels can vary considerably between different sectors of the industry.

The *profit* can be expressed as the difference between the contract sum and that required to pay for overheads, site labour costs, materials and plant hire costs needed to complete the project.

The profit included by a contractor is influenced by many factors, which can be collectively referred to as 'market conditions'. The following is a short list of the principal factors to be considered by the contractor prior to establishing a margin of profit in its estimate:

- Quantity of work currently in hand
- Future commitments
- Placing in previous tenders
- Possibility of future work from the same source
- Assessment of the degree of competition.

An example of a profit allowance calculation is shown in the following example.

**Typical *profit allowance* calculation**

|  | £ p |
|---|---|
| Gross turnover | 2,000,000.00 |
| *Less* profit | 100,000.00 |
|  | 1,900,000.00 |
| *Less* general overheads | 250,000.00 |
| Net annual cost | 1,650,000.00 |

Net profit as a percentage of net cost (turnover)

$$\frac{100,000.00}{1,650,000.00} \times 100 = 6 \text{ per cent}$$

# Synthesis of unit rates

The synthesis of unit rates requires the estimator to add together the costs of all of the labour, materials and plant needed to complete a given item of work. The estimator must have the necessary skills and detailed knowledge needed to identify all the various items to be included in any particular unit rate.

The technique also involves the estimator converting the individual elements of the unit rate into the correct unit for the particular measured items ($m^3$, $m^2$, m, No. or Item).

# Pricing policy

All contracting firms have a pricing policy when preparing a unit rate estimate. The pricing policy will normally be based on one of two main techniques:

- Net unit rates
- Gross unit rates.

The technique selected will depend on whether the contractor makes an allowance for overheads and profit in the unit rate build-up.

### Net unit rate

This is a policy where the contractor makes no allowance in the calculation for the overheads and profit. The overheads and profit are normally assessed separately and then added to the estimator's total net cost for the whole project so as to give the final tender figure. An example of a calculation for net unit rate is shown in the example.

## Typical calculation of *net unit rate*

| Code 3 lead flashing 150 mm girth | £ p |
|---|---|

*Material*
Cost of Code 3 lead delivered
to site per tonne ..................................... 1,100.00

Allow 2 hours per tonne to unload and
stack 2 hrs @ £7.85 ................................... 15.70
                                                     ─────────
                                                     1,115.70
*Add* 7.5% for waste and laps ......................... 83.68
                                                     ─────────
                                                     1,199.38

If each m$^2$ uses 14.18 kg of Code 3 lead
then
Total area per tonne of Code 3 lead

$$\frac{1000\,\text{kg}}{14.18} = 70.53\,\text{m}^2$$

Therefore total length of flashing 150 mm wide per tonne

$$\frac{70.52}{0.150} = 470.20\,\text{m}$$

Therefore cost per m

$$\frac{£1,199.38}{470.20} = £2.55 \qquad\qquad 2.55$$

No allowance for off-cuts for wedges and tacks
(included in waste and laps)

*Add* allowance for mortar for pointing (say) ............ 0.16

Labour based on gang cost of plumber and
mate at £7.85 + 6.64 = £14.49 per hour
Labour constant at 5 m per hour

$$\frac{£14.49}{5} = 2.90 \qquad\qquad 2.90$$

**Net unit rate** .......................................... £5.61

## Gross unit rate

The prices calculated by the estimator include some predetermined percentage allowance for the overheads and profit. An example of a calculation of gross unit rate is shown below.

## Typical calculation of *gross unit rate*

| Code 3 lead flashing 150 mm girth | £ p |
|---|---|
| Carried forward from previous calculation | 5.61 |
| *Add*<br>Allowance for head office overheads @ 16% | 0.90 |
| | 6.51 |
| *Add*<br>Allowance for main contractor's profit @ 6% | 0.39 |
| **Gross unit rate** | **£6.90** |

# Sources of information

Accurate and reliable estimating depends upon the collection, analysis, publication and retrieval of pricing information. The principal sources of estimating data include:

- Bills of Quantities
- Price books
- Technical journals.

## Bills of Quantities

The contractor keeps on file previously-priced Bills of Quantities as a useful source for pricing new tenders. These records are particularly useful if the new project being priced is of a similar nature to the previous project.

## Price books

Price books are published at the beginning of each year. Prices change each year so the sources gradually become out-of-date. The layout and presentation of price books vary, but the following information is usually included:

- Calculation of basic wage rates
- Daywork rates
- Professional fee for consultants
- Market prices for materials
- Labour constants
- Prices for measured work
- Approximate estimating section.

Some well-known price books in the United Kingdom include the following:

- *Spon's Architects and Builders Price Book*
- *Laxton's Building Price Book*
- *Griffiths Building Price Book*
- *Hutchins' Priced Schedules.*

### Technical journals

Many periodicals frequently contain supplements on building costs. Popular titles in the United Kingdom include the following:

- *Building*
- *Building Trades Journal*
- *QS Monthly*
- *Architects' Journal.*

### Other sources of information

Other popular sources of information for the estimator include:

- Subcontractors and suppliers
- In-house company information
- Government literature
- Personal information.

---

### Key words for pricing

The following is a list of some keywords used in this chapter. Use the list to test your knowledge and, if necessary, consult the text to learn about the terms.

| | | |
|---|---|---|
| All-in rate | Synthesis | Measured net |
| Labour constant | Net unit rate | Discounting |
| Overheads | Profit | Conversion factors |

# Index